Unfolding Mathematics with Unit Origami

BETSY FRANCO

DRAWINGS BY DIANE VARNER

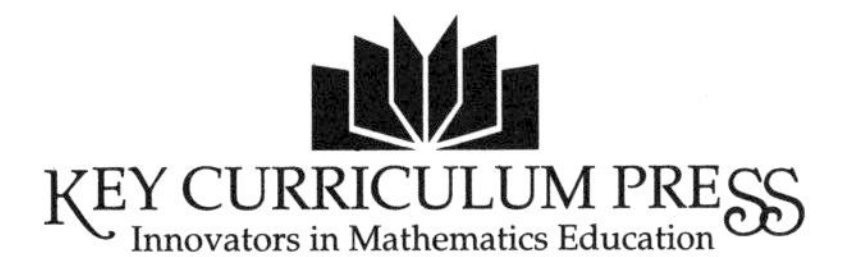

Editorial Development: Sarah Block, Casey FitzSimons, Greer Lleuad, Crystal Mills
Editorial Assistance: Caroline Ayres, James A. Browne, Romy Snyder
Contributors: Vernon Isaac, Laura Kruskal, Robert Lang, Jeremy Shafer
Consultants: Rona Gurkewitz, Joe Malkevitch, David Masunaga, Jeanne Shimizu-Yost
Production Editor: Deborah Cogan
Copyeditor: Rosana Francescato
Production Manager: Diana Jean Parks
Production Consultant: Steve Rogers
Design Coordinator: Diana Krevsky
Cover and Interior Designer: Carolyn Deacy Design
Art Editor: Kelly Murphy
Illustrator: Diane Varner, Diane Varner Illustration
Technical Illustrator: Ben Turner Graphics
Photographer: Ann Dowie
Origami Models: Crystal Mills, Wei Zhang
Page Layout: Kirk Mills
Production Assistance: Michelle Kolota, Ann Rothenbuhler
Cover Photographer: Hillary Turner

Editorial Director: John Bergez
Publisher: Steven Rasmussen

LIMITED REPRODUCTION PERMISSION

Key Curriculum Press
1150 65th Street
Emeryville, CA 94608
510-595-7000
editorial@keypress.com
http://www.keypress.com

Printed in the United States of America 10 9 8 7 6 5 4 3 2 02 01 00 99 98

ISBN 1-55953-275-0

Contents

Preface v

Overview vii

- Unit Origami, Mathematics, and the Classroom vii
- How to Use This Book vii
- Practical Classroom Tips ix

Key to Diagrams xiv

The Art of Making a Fold xv

Activity 1 Getting Started 1

Activity 2 Folding a Square I 5

Activity 3 Folding a Square II 12

■ *My Experiences with Origami* by Vernon Isaac 19

Activity 4 Ori Money in Two Dimensions 21

Activity 5 Triangular Measuring Tool 26

Activity 6 Folding a Regular Hexagon 32

Activity 7 Free Folding an Equilateral Triangle 38

■ *My Experiences with Origami* by Jeremy Shafer 43

Activity 8 Folding a Regular Pentagon 45

Activity 9 Octagon Star 52

Activity 10 Star-Building Unit 57

Activity 11 Triangular Hexahedron 65

■ *My Experiences with Origami* by Robert J. Lang 73

Activity 12 Cube 75

Activity 13 Two Colliding Cubes 82

Activity 14 Stellated Octahedron 89

Activity 15 Stellated Icosahedron 96

▪ *My Experiences with Origami* by Laura Kruskal 102

Activity 16 Free Folding More Polyhedra from Star-Building Units 104

Glossary 109

Bibliography and Resources 113

Preface

In *Unfolding Mathematics with Unit Origami*'s sixteen activities, you and your students will be introduced to some origami history, the basic mathematical ideas related to origami folding sequences, and some exciting three-dimensional unit origami models. The activities can be used with geometry and algebra classes, and are appropriate for ninth- and tenth-grade students as well as middle school students. Additionally, the activities offer opportunities for group work, journal writing, and projects. Autobiographical vignettes throughout the book may prompt students to discuss their own experiences with origami. The overview that follows explains in detail how you can best use this book in your classroom.

Although we usually associate origami with Japan, it is an art and a craft that is enjoyed by people throughout the world. Origami designs that originate in different countries often have their own distinctive characteristics. There are many local and international organizations in which people get together to share new models or to work together to learn how to fold models. It is truly a unique experience to meet with a group of international folders. Even if you don't speak a common language, you can communicate with an origami folder through the "peace" of paper. Many people who enjoy origami regard it as a sociable pastime through which they can meet with others to share ideas. A meeting of origami enthusiasts exemplifies cooperative learning at its best! Origami knows no age boundaries. Very young children can learn how to fold origami models, and some children even learn how to follow origami folding diagrams before they learn to read. One of the leading origami masters in Japan, Akira Yoshizawa, is in his nineties and still folding. It is said that he has created over fifty thousand original models.

Many people who do origami enjoy following diagrams that others have created. Other origami folders are intrigued by designing original figures and models. Some designers envision what they want to create through an entirely creative process. It's not unusual for an origami designer to get inspiration for a new folding sequence while sleeping. Other origami designers use a more mathematical approach, sometimes creating folding sequences on graph paper or with the help of computer programs. Origami is rewarding for people from a wide variety of backgrounds and occupations, and what people do with origami is as varied as the people doing it. Once you get hooked on origami, the possibilities for where it will lead you are unlimited.

The models in this book are based on the concepts of unit, or modular, origami. In unit origami, a geometric figure is folded from one sheet of paper. The folding sequence used to fold the figure is repeated with several more sheets of paper to produce a collection of identical units, or modules. These units fit together to form a model. The model might be a polyhedron such as a cube, a dodecahedron, or an icosahedron, or it might be a two-dimensional star, a box with a lid, or some other decorative shape. Unit origami is a relatively new type of origami. The premier designer of modular folds is Tomoko Fusè of Japan. Ms. Fusè, a horticulturist who also cultivates an interest in origami, has written many books and has created many original unit origami models.

In this book we have borrowed ideas from many different sources. We have tried to give credit where credit is due; if we have failed to do so, please let us know. We are especially grateful to Tomoko Fusè for the work she has done with unit origami. Robert Neale was the originator of the octagon star, which he calls the *Pinwheel-Ring-Pinwheel.* The dollar bill folding sequence is based on ideas written about by Jean Pedersen in a letter to the editor of *Mathematics Magazine* 61, no. 4 (October 1988): 270. David Masunaga was an invaluable resource and shared with us projects and ideas from his own experiences working in the classroom. Jeanne Shimizu-Yost also shared many of her experiences with teaching the star-building unit.

We hope that you and your students will enjoy the origami activities in this book and that your students will appreciate the mathematics associated with paper folding. Relax, have fun, and enjoy. Perhaps this experience will be the beginning of a lifelong adventure in paper folding.

Overview

UNIT ORIGAMI, MATHEMATICS, AND THE CLASSROOM

Implementing unit origami in the classroom can mean rich, hands-on geometry experiences for your students. The materials are simple, the mathematics is pervasive, and the end product is satisfying and beautiful. Students involved in origami have the opportunity to use the vocabulary of mathematics in context and to discuss geometric concepts as they assemble origami models. As you will see in the activities in this book, there are innumerable mathematical skills and concepts related to origami. Here are some of these skills and concepts:

- Spatial visualization
- Regular polygons
- Triangles
- Congruence and similarity
- Fractions and ratios
- Sequences
- Angles
- Regular solids
- Semiregular solids
- Dual polyhedra
- Stellated polyhedra
- Inscribed polyhedra
- Deltahedra
- Symmetry
- Area, volume, and surface area
- Cross sections
- Pythagorean theorem
- Euler's formula

HOW TO USE THIS BOOK

How This Book Is Organized

This book is divided into sixteen activities that both explore the mathematics of the featured models and explain how to fold the models. Activities 1–7 introduce students to origami and provide folding practice. In Activities 8–16, students learn to fold a series of polyhedra. In Activities 7 and 16, students are encouraged to explore and extend in an open-ended way the concepts they have just learned.

For each activity, the student pages are directly followed by the teacher pages. Most of the student activities include these sections:

- Illustration of the completed model
- Introduction to the activity
- List of materials needed
- Suggestions for student groupings
- Folding or assembly instructions with questions
- Questions for exploring your model

In addition, in the Teacher Notes you will find the instructions needed to complete an activity and suggestions for extensions of the activity.

Other Features in This Book

Throughout the book you will find vignettes about people who enjoy doing origami. You may want to read these vignettes to your class or ask students to read them. Students may enjoy finding out more about what kinds of people enjoy doing origami and how they got started doing it.

A glossary of mathematical terms is provided for your convenience. It is probably best not to pass out the glossary to students, because the exploration questions in the activities often ask students to define for themselves a concept that is defined in the glossary.

Where to Begin

The first activity in the book, "Getting Started," has background information on traditional origami, an explanation of unit origami, and tips for preparing your mind and body for origami. Sharing this information with your students can set the tone for successful paper folding.

The two sequences of activities have been carefully prepared in terms of increasing difficulty of both folds and mathematical concepts. You can jump in anywhere in the book, but if you do the activities in order you can cover a broad range of mathematics topics.

Trying out the folds at home before introducing them in class will make you feel more secure in the classroom. The book starts with two-dimensional geometric models so that the folds remain relatively simple while you and your students get familiar with the origami directions and gain confidence in your paper folding.

Teaching an Activity

Each activity gives you teaching ideas and strategies. Here is one way to present an activity:

1. Designate partners or groups so that students can help each other and exchange ideas.
2. Use the introductory paragraph as motivational material to begin or prepare for the activity.
3. Distribute photocopies of student pages and other materials.
4. Demonstrate folds using large squares of paper that all the students can see, or on the overhead projector using waxed paper as students follow along.
5. Have students respond orally to the questions asked throughout the folding instructions. Sometimes you may ask students to record their answers in their origami journals.
6. Circulate around the room, helping groups and demonstrating how to assemble units.

7. Have students answer the "Exploring Your Model" questions in their origami journals.
8. Discuss the "Exploring Your Model" questions as a class and reflect on the activity.

PRACTICAL CLASSROOM TIPS

What Supplies Will I Need?

There are several options for paper:

- Origami paper (supplied through Key Curriculum Press or found at a local art store).
- Colored ditto or mimeograph paper, or notebook paper. One sheet can be cut into one $8\frac{1}{2}$-by-$8\frac{1}{2}$-inch square or two $5\frac{1}{2}$-by-$5\frac{1}{2}$-inch squares.
- Wrapping paper or pages from old magazines cut into squares.
- Patty paper, the paper used to separate hamburger patties, can be purchased from Key Curriculum Press or obtained at restaurant supply stores. This is the least expensive option. It is a good medium for students to use when they are learning how to do a new fold, and for the first few activities when they are experimenting with different folds.
- Fadeless® brand bulletin paper is readily available at most school sites. It comes in rolls or flat sheets (unfortunately not square) in various widths from 12 to 48 inches. Flat sheets are easier to cut and handle but come only in 12- and 18-inch widths.
- Brightly colored paper can be found at photocopy centers.

(See the "How Can I Teach My Students?" tips in this section for supply options for demonstrations.)

What Size Squares Should I Cut?

Six-inch squares are optimal, but larger squares may be necessary for students whose fine motor skills are not well developed.

How Much Time Should I Set Aside?

Unit origami involves time for demonstration, for folding multiple units, and for assembling models. Most of the activities in this book are designed to take one class period. Some teachers introduce the basic unit folds in class and assign the rest of the folding as homework. In any case, students will get the most out of origami if they are careful and thoughtful about their folding, which means that they will need to slow down enough to be precise.

Leaving enough time for the first few activities is important, because it takes time for students to become comfortable with the folds and to appreciate origami on both a mathematical and an artistic level.

What Preparation Is Required?

You will need to fold each model at home first so that you are fairly confident about its construction. However, don't wait until you feel like an expert to get started. You can learn along with your students, some of whom will become so adept that they can be your assistants. You may even want to ask students to present the folding instructions to the class.

While you are learning how to fold a particular model, you can think about the mathematical concepts involved in its construction and how these fit into your curriculum. This is also a good time to consider how much paper you will need for the project and how you will group your students.

What Is an Effective Way to Group My Students?

Origami requires patience and can be a bit intimidating for a student to tackle alone. This is especially true for more complex models, although most of the models in this book are quite simple. Also, dealing with the number of squares needed can make construction of individual student models much more time-consuming. Grouping suggestions are given for each activity. For many activities, pairing students and then forming groups of four is the optimal arrangement. This enables students to discuss and interpret the instructions together. Also, assembling models is often easier with four hands. Note that each partner can construct his or her own model, or a pair can construct a shared model. For more complicated models and those involving many folded units, students are encouraged to assemble one model per group.

How Can I Teach My Students?

To guide students through the activities, you'll need to both demonstrate how to fold the models and coach students as they explore the activities on their own. The methods for your presentation will be different depending on the size of the group.

Large groups

- Use an overhead projector and waxed paper. The paper is translucent and the folds are easy to see when projected.
- Use a large piece of paper. Twenty-four inches is a good demonstration size. A black marker can be used to highlight the folds.

The table at the top of the next page outlines the advantages and disadvantages of these two methods.

Small groups and individuals

- Use the same size square as your students.
- Display a poster showing a sample of each individual step.
- Give the students a copy of the diagrams and possibly a sample of the final product. Let them figure it out.

Method	Advantages	Disadvantages
Overhead	More people can see. Teacher is always facing the class. Waxed paper is readily available.	Teacher can get stuck at overhead projector. Overhead projector might not be available. Overhead projector shows only two dimensions. Waxed paper doesn't show colors. Waxed paper isn't good for demonstrating assembly of units.
Large paper	Paper is colorful. The product is three-dimensional. The product looks like that of the students.	Often teacher's back is to the class. It requires more coordination and expertise to demonstrate and face the class. Large squares must be cut by teacher. Paper might not be available.

The process

Start by demonstrating the folds slowly, step by step. Liberally incorporate the language of geometry into the visual demonstration. At the end of your demonstration, those who have the idea can go on to make the number of duplicate units required by the model. Meanwhile, you can demonstrate one more time, at a faster pace.

If you are not assigning the rest of the folding for homework, your role switches at this point. You can begin circulating around the room, listening in as students teach each other and intervening only when necessary. Students whose motor skills and spatial visualization are not advanced may need extra help at this time.

Some students may be ready to assemble the units. As you circulate, you can teach the fastest person in each foursome how to assemble the units. That person then becomes the teacher for the rest of the foursome. An alternative is to train leaders beforehand and put one leader in each foursome. If a group of four is having difficulty, their neighboring group can assist them. Origami is fertile ground for cooperative learning!

What Do I Do If a Model Requires Many Units?

There are several alternatives if a model requires twelve, fifteen, or even thirty units. Groups can form an assembly line to hasten the folding process. Squares can be taken home and brought back to class. Or each group can make one model.

How Can I Store Unfinished Models?

The easiest way to keep track of units that haven't been assembled and models that are unfinished is to store materials in resealable plastic bags. Students can write their names and the name of the polyhedron they are building right on the bag, using a permanent pen. Neatly storing these bags in a box or a cupboard keeps them organized and safe.

What Can I Do with the Completed Models?

Completed origami models can be displayed around the room and hung from the ceiling, individually or as mobiles. In fact, they can serve as motivators, and as finished samples, for next year's students. They can also be exhibited in a school art show or display case. Eventually, they can be taken home or given as gifts.

Is Origami Appropriate for Homework?

Folding multiple units can easily be done at home, and the assembly can take place in the classroom. The free-folding activities and the extensions can also be completed as homework. Unlike so many manipulatives that are found only at schools, paper is readily available.

How Do I Explain the Noise Coming from My Classroom?

There is a difference between noise that is social chatter and noise that results from discovery, investigation, experimentation, and an exchange of ideas. Origami makes geometry come alive through hands-on activity and purposeful discussion.

Writing About Mathematics and Origami

Most of all we enjoyed watching things and shapes come to life from a piece of paper.

—Middle school students

Throughout this book there are opportunities for students to write about their experiences with origami and geometry. Students can record their discoveries about geometry in their origami journals; can respond to the vignettes, described below, in each sequence; and can write up the results of group projects.

Origami Journals

Keeping an origami journal will help students stay organized, encourage writing, and enable students to record connections they see between geometric ideas. In each activity, a set of exploration questions appears after the folding or assembly instructions. Generally, these questions present problems that require in-depth thinking and experimentation. The answers formulated by pairs or groups of students can be recorded in students' origami journals.

Vignettes

Vignettes about people who enjoy doing origami appear throughout the book. You might want to give students an opportunity to respond to these vignettes on a personal level. The art of paper folding and its mathematically exciting results have the power to boost self-esteem, provoke thoughts and feelings, teach students about themselves, and allow them to reflect on their learning processes.

Projects

On occasion, you might want to approach an activity, or part of an activity, as a group project to be written up and presented. This can happen in several ways:

- Free-folding activities at the end of each sequence can be used for group projects.
- Activity extensions that involve the folding or assembling of polyhedra can be used for group projects.
- Groups can choose an activity from a sequence to use for a group project.

The write-up of a group project can include some or all of these elements:

- Description: a description of the origami figure you are making and the materials needed
- Purpose/Objectives: the reasons you picked this figure and what you expect to learn
- Instructions: your own step-by-step directions for folding the figure
- What You Learned: a description of what your group learned about geometry, about origami, and about working together

How Do I Assess Origami Work?

Here are two grading rubrics that teachers have used:

Model	30%
Mathematics	30%
Folding instructions	20%
Design (layout of poster, etc.)	10%
Neatness	10%
Extra credit	5%

Completeness of model	0	1	2	3	4	5
Writing about mathematics	0	1	2	3	4	5
Folding instructions	0	1	2	3	4	5
Design of display	0	1	2	3	4	5
Neatness	0	1	2	3	4	5
Extra credit	0	1	2	3	4	5

Where Can I Learn More About Origami?

This book is likely to keep you busy for a while, but many other origami books are available that describe different models and offer different perspectives. Other sources are listed in "Bibliography and Resources" at the end of the book.

Key to Diagrams

Arrows and dotted lines are the main ingredients of origami directions. You can use this page of directions as a handy visual guide.

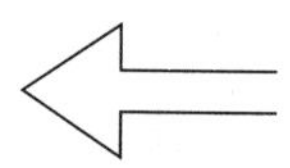

Pull or open out.

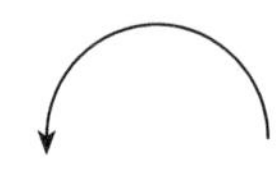

Rotate the paper.

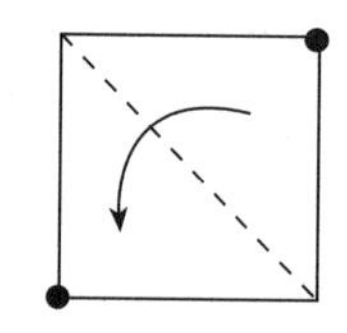

Fold so the dots meet.

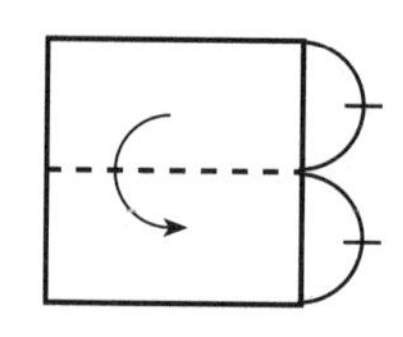

Fold exactly in half.

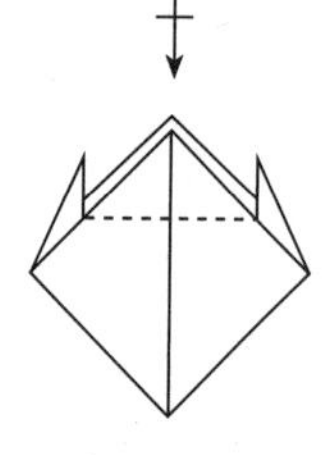

Repeat on the underside.

Fold and unfold.

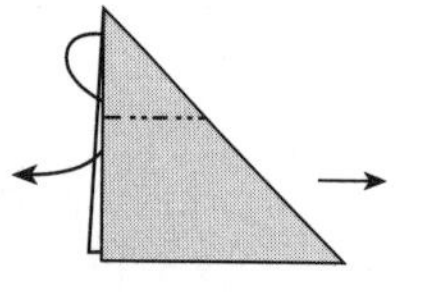

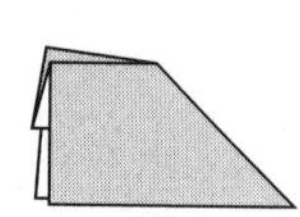

Fold inside by reversing the fold.

The Art of Making a Fold

Folding a piece of paper sounds like an easy thing to do, but origami can be more enjoyable if you take the time to make a careful fold. Take a few minutes to look at these steps, which show how to fold a piece of paper in half:

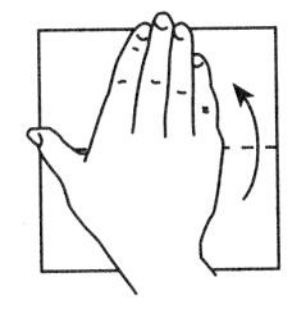

Work on a flat surface. Sweep your fingers over the paper to flatten it.

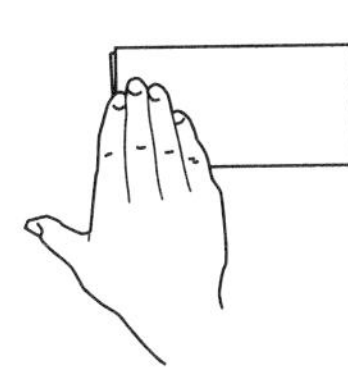

Match up the edges of the paper.

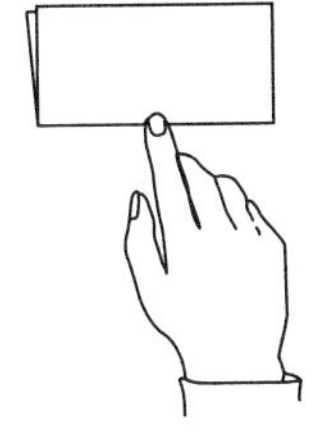

Make a light preliminary fold in the middle of the paper.

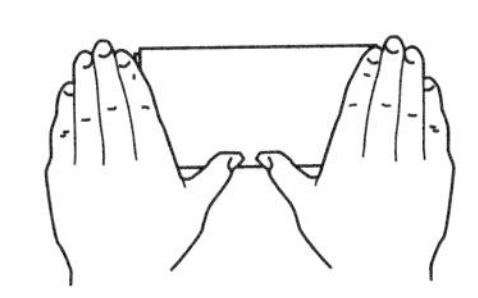

Check the edges of the paper and adjust if necessary.

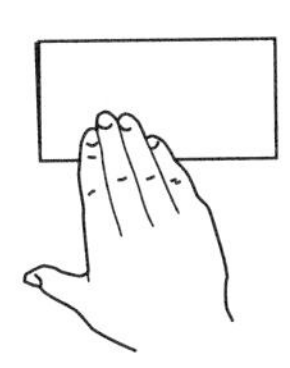

Follow up with a strong crease, using your fingernail if necessary.

If you make a careful fold, each fold that comes after it will be easier. Keep these guidelines in mind when folding:

- Looking at the picture of the result of the fold can help you understand the fold.
- There is usually only one possible way to do a fold. Finding that way may require some experimentation.
- Handling the paper too gingerly will result in creases that are not sharp or well-defined.
- The colored and uncolored sides of the paper can help you get oriented in relation to the diagram.
- Extraneous fold lines that you make on the paper may be confusing. If you have experimented so much that you have a lot of extra fold lines on the paper, you might want to start with a fresh piece of paper.

Unfolding Mathematics with Unit Origami

Getting Started

An origami design springs to fresh life each time someone executes it and in this sense may be regarded as eternal.

Tomoko Fusè

TRADITIONAL ORIGAMI

The word *origami* means "paper folding" (*ori,* "to fold"; *kami* or *gami,* "paper"), but this literal translation does not tell the whole story. The Japanese word *kami* is also a homonym for the word meaning *"god,"* hinting at the deeper meaning of origami to the Japanese people.

折り紙

Origami written in Japanese characters

Many people have had some experience with traditional origami, which involves folding one uncut piece of paper into an object. The paper crane is recognized universally.

It is believed that origami started over one thousand years ago, sometime after paper was first introduced in Japan. Because paper was very expensive in those days, origami could be used only for ceremonies, not for pure fun and enjoyment. Besides, the folds were made in such a formal way that they were very difficult to learn.

In the late nineteenth century, a paper dealer in Tokyo made an important decision for origami. He began importing paper from Europe and cutting it into squares, rather than into rectangles as had been done in the past. Then he sold the colored squares as origami paper. This was the beginning of the origami that is popular today.

Eventually, origami became a part of everyday Japanese life. In fact, it became a craft for children. Parents and grandparents taught it to their five- and six-year-old children—and they still do. In this way, origami has been handed down for centuries.

Origami plays an important part in certain Japanese festivals. For Tanabata (the Star Festival) on July 7 each year, origami decorations are hung on a branch of bamboo, along with wishes written on narrow strips of paper. On May 5, which used to be Boys' Day and is now Children's Day, children use large pieces of paper to fold Samurai helmets.

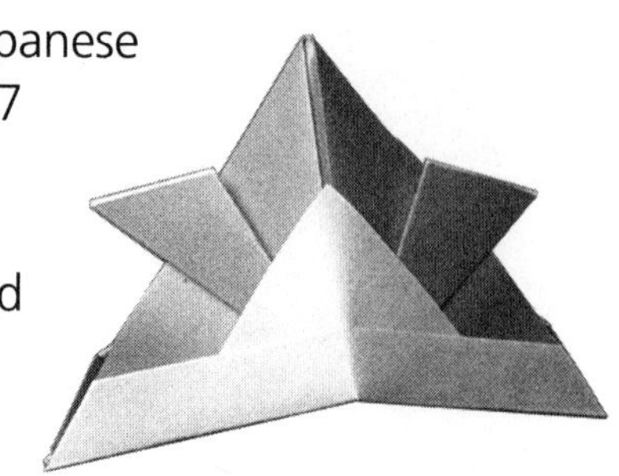

QUESTIONS

1. What is your experience with traditional origami?
2. Are any traditions and skills passed down in your family? If so, what are they?

WHAT IS UNIT ORIGAMI?

Unit or *modular origami* is a specialized type of paper folding that breaks with tradition by allowing the folder to use more than one square of paper to form geometric objects called *models.* Multiple copies of the same unit are folded to create the models, which are assembled using built-in pockets, as well as independent joints, usually without any glue or adhesive.

Kunihiko Kasahara, who created some of the folds described in this book, and Tomoko Fusè have been very influential in the field of unit origami. Their creations are exciting and beautiful. When describing unit origami, Fusè says that it has the intrigue and fascination of a puzzle. She also speaks of its surprising and incalculable quality.

One of the most intriguing aspects of unit origami is that it is a relatively new field and there is much territory yet to be explored.

QUESTIONS

1. What have you learned from puzzles?
2. What do you expect to learn by folding origami figures?

WHO DESIGNS, CREATES, AND FOLDS ORIGAMI MODELS?

Some people think that all origami designers and folders are from Japan. Believe it or not, there are organizations all over the world for people who are interested in origami. If you ever get the opportunity to attend an origami convention, or even a local folders' group in your area, you'll meet people of all ages. Some children learn to fold before they learn to read. You will meet people from all professions who somehow got interested in origami. Many people find origami a creative and relaxing pastime. To find out more about some people who enjoy doing origami, you can read the vignettes in this book.

Origami diagrams are written in a universal language so that people who don't speak the same language can still enjoy folding an origami model together. Origami enthusiasts often refer to origami and the "peace" of paper.

Origami designers and creators are usually happy to share their creations with other folders, but it's important to give credit to a creator when you share his or her design with someone else. Whenever possible, the name of the person who created a folding sequence has been included in these activities. As you discover other sources for origami diagrams and models, always pay attention to who created the models. Soon you may even be able to identify a certain model or diagramming technique as belonging to a certain creator.

1. Describe an activity you do with a group of people in which different age groups participate.
2. Describe an experience you might have had with someone who didn't speak your language. How did you communicate?

PREPARATION OF MIND AND BODY

Understanding something intellectually and knowing the same thing tactilely are very different experiences.

Tomoko Fusè

When you are ready to start unit origami, you will need to keep in mind that it is best to start folding models right away. Just flipping through the diagrams can be confusing. Origami is a hands-on craft that *anyone* can do. The understanding of how to fold a piece comes only from working it out step by step. It requires an open mind and the patience to investigate the folds until you discover the key to the folding and assembly. Inevitably, the results are deeply satisfying. Fusè admits that her own first versions are wrinkled and messy. She encourages the folder to persevere.

QUESTIONS

1. What other things do you do that require patience and the ability to stick to something until you figure it out?
2. Describe one of these activities and explain why it is worth your time and patience.

Getting Started

OBJECTIVES

To introduce students to the history and art of origami
To appreciate the traditional aspects of origami
To be introduced to the connection between puzzles and origami
To think about the importance of patience and perseverance

MATERIALS NEEDED FOR EACH STUDENT

Origami journal
Student worksheet pages

TEACHER MATERIALS

None

TIME

Half a class period

GROUPING

The whole class will work together.

GENERAL INSTRUCTIONS

In this lesson students are introduced to the history and tradition of origami. Students think about experiences they have had with puzzles and how some activities require much patience. You could introduce the idea of an origami journal in this activity and ask students to answer the questions in their journals.

ACTIVITY 2

Folding a Square I

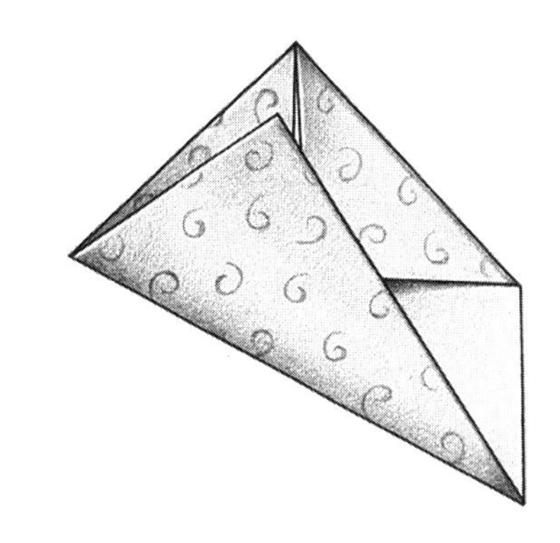

The word *origami* means "paper folding." The basic tool of origami is a simple square. In this activity you will fold a square many different ways and begin to see the unlimited possibilities that the square has to offer.

You will start by exploring different ways you can fold a square in half. While doing this, you will relate the results to different polygons. This activity will help you better understand what's happening when you learn to do more complicated origami folds.

Many people think the art of paper folding requires special origami paper, but you can use any kind of paper that is available. Patty paper works very well for this activity.

MATERIALS NEEDED FOR EACH STUDENT

At least eight squares of origami or patty paper

Origami journal

GROUPING

Work in groups of four.

FOLDING A SQUARE IN HALF

1. How many different polygons can you fold where each has an area that is one-half that of the original square? Make as many different polygons as you can.

The original square.

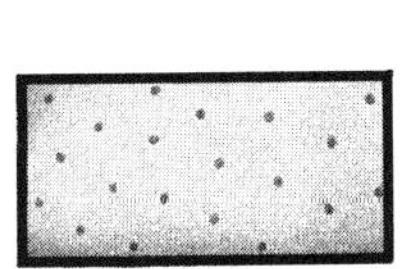

The area of this polygon is one-half of the original square.

The area of this polygon is greater than one-half of the original square.

2. Work with your group to determine different ways you could classify your results to question 1. Be prepared to present your classification method to the class.

3. In a regular polygon all of the sides' lengths are equal and all of the interior angles have the same measure. Which of the polygons that you created in question 1 are regular polygons? Create a chart like the one below in your origami journal. Add to your chart any regular polygons you have encountered so far. You will fill in more information as you complete other origami activities.

Name of regular polygon	Number of sides (angles)	Number of axes of symmetry	Kind(s) of rotational symmetry

4. How can you tell when you've found all the possible ways to fold a square in half? Or are there an infinite number of possibilities? Write about your conjectures.

5. Look at the fold pictured at the right. Has the square been folded in half? If you think so, provide a convincing argument.

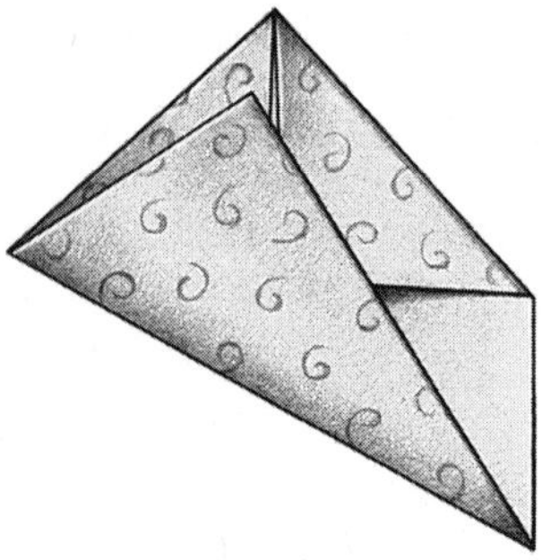

SQUARE SYMMETRIES

Write your answers in your origami journal.

1. If you can fold a figure on a line so that both halves of the figure lie on top of each other, then the figure has reflection symmetry. Some of the fold lines you made when folding your square in half are lines of reflection symmetry. Take a new square and fold the square in half as many times as you need to find all the lines of reflection symmetry. Unfold your square, and make a sketch in your origami journal showing these lines of reflection symmetry.

   a. How many lines of reflection symmetry does a square have? Describe these lines of symmetry.

b. Fold a square along any line of reflection symmetry. What is the area of the new polygon compared to the area of the original square? Write a conjecture based on this result about a line of reflection symmetry.

2. If a figure can be rotated about a point in such a way that its rotated image coincides with the original figure after turning it less than 360°, then the figure has rotational symmetry. The letter Z has 2-fold rotational symmetry because when it is rotated 180° and 360° about a center of rotation, the image coincides twice with the original figure.

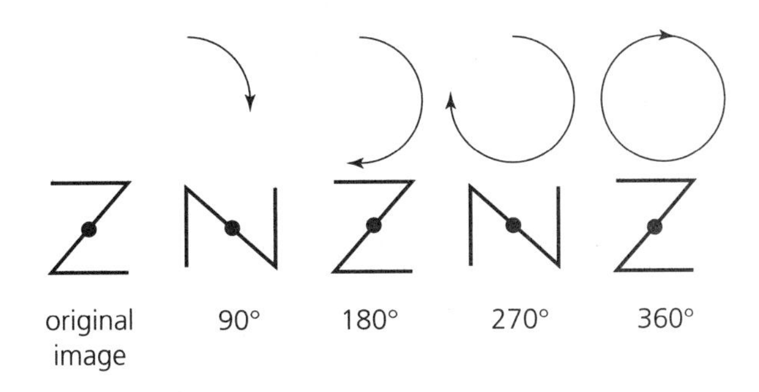

a. Trace your square on another piece of paper. Label the upper left-hand corner of your square paper A. Place the square paper on top of the traced square. Now rotate the square paper 90° clockwise. Count how many times you have to rotate it 90° until the A is in the left-hand corner again.

b. Describe the rotational symmetry of the square.

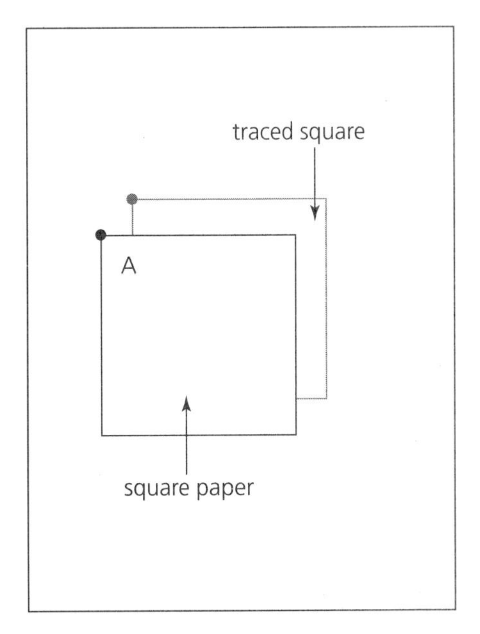

ACTIVITY 2

Folding a Square I

OBJECTIVES

To fold a square in half
To begin a list of regular polygons and their characteristics
To explore the reflection and rotational symmetry of the square
To appreciate the versatility of a square piece of paper

MATERIALS NEEDED FOR EACH STUDENT

At least eight squares of origami or patty paper
Student worksheet pages—one copy for every two students
Origami journal

TEACHER MATERIALS

Overhead projector and waxed paper, or large paper for demonstration

TIME

One class period

GROUPING

Students should work in groups of four. Each student will do his or her own folding.

GENERAL INSTRUCTIONS

Folding a square in half is a nonthreatening way to introduce paper folding. Students will feel successful and gain confidence as they find interesting ways to fold.

This activity is designed to be very open-ended. Encourage students to do lots of exploring when they're folding a square in half. This activity should lead to some interesting questions and discussion. Making and displaying class charts showing results will help students to see the almost unlimited possibilities. The Square Symmetries section could be assigned as homework and reviewed the next day in class.

Patty paper works very well for this activity. Be sure to have an ample supply so that students can feel free to try many ideas and methods when experimenting with folding squares in half.

Be sure to encourage students to record their results in their origami journal. Making sketches of the resulting figures will help students interpret diagrams in later activities.

ANSWERS

Folding a Square in Half

1–3. Answers will vary for student entries in the charts.

While working in groups, students can keep a class record of ways to fold a square in half by tacking their discoveries on the bulletin board. Only those discoveries that are distinct should be part of the display. The first student to discover a method can write his or her name on the figure. The following are some different ways to fold a square in half.

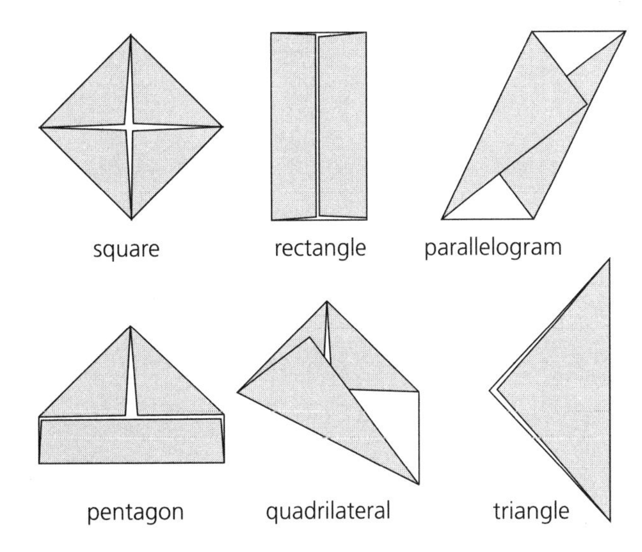

The folded parallelogram in the preceding figure involves the upper right vertex and the lower left vertex of the square. Because there are two pairs of opposite vertices on the square, there are actually two ways to fold the parallelogram. The results are mirror images of each other, representing distinct ways (shown at right) of folding a square in half.

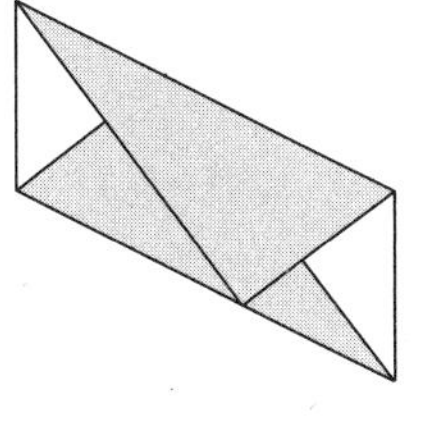

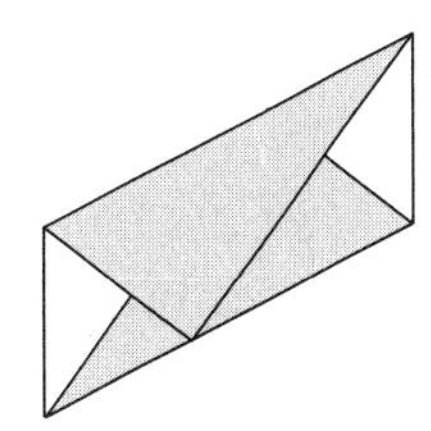

Some of these folds are so common in origami that they have been given names. The first fold shown in the six-item figure above is called a *blintz fold,* and the second fold is called a *cupboard fold.* The pentagon shows a combination of a *house roof fold* and half of a *cupboard fold.* The triangle is called a *diaper fold.* You may want to introduce these terms to your students. You could add these names to the figures displayed on the bulletin board.

2. One way to classify the results is to count the number of folds in the completed model. You could create a chart like the one below and have students pin each of their polygons in the appropriate cell. You could also ask students to draw sketches showing their polygons.

One fold	**Two folds**	**Three folds**	**More than three folds**

Another way to classify the results is to look at the type of polygon formed. You could create another chart like the one below.

Polygon			
Triangle			
Square			
Rectangle			
Parallelogram			
Quadrilateral			
Trapezoid			

3. The only regular polygon that has an area equal to one-half that of the original square is the smaller square. Students should add the square to their chart.

4. Students will probably come to the conclusion that there are an infinite number of ways to fold a square in half.
5. Students will come up with different methods for showing that the area of the quadrilateral is half that of the original square. One possibility is cutting the paper (or tracing the shapes) and showing that both halves are equal in area. If you turn the quadrilateral over, you can easily see that its area is one-half that of the original square. The diagram at the right shows how to fold the quadrilateral.

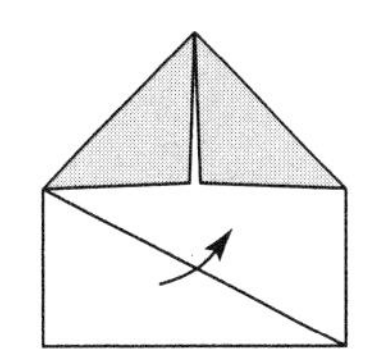

Square Symmetries

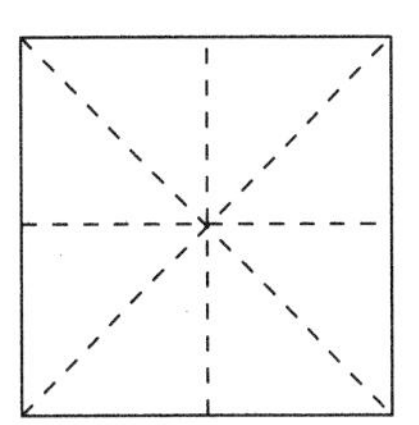

1. Students should make a sketch that looks like the one shown at the right.
   a. Four lines: Two of the lines of reflection symmetry are diagonals, and the other two are perpendicular bisectors of the sides.
   b. Whenever you fold a square along a line of symmetry, the area of the resulting polygon is half that of the original square. Lines of reflection symmetry cut a square in half.
2. a. Four times.
   b. The square has 4-fold and 2-fold rotational symmetry. (Note that if a figure has x-fold rotational symmetry, it will also have "any factor of x"-fold rotational symmetry. For example, if a figure has 6-fold rotational symmetry, it will also have 3-fold and 2-fold rotational symmetry.)

ACTIVITY 3

Folding a Square II

In this activity you will investigate folding a square into many parts. You will also learn how you can make line segments of different lengths. And you will use some geometric patterns to discover some sums of sequences. Perhaps you will begin to appreciate how versatile a square really is.

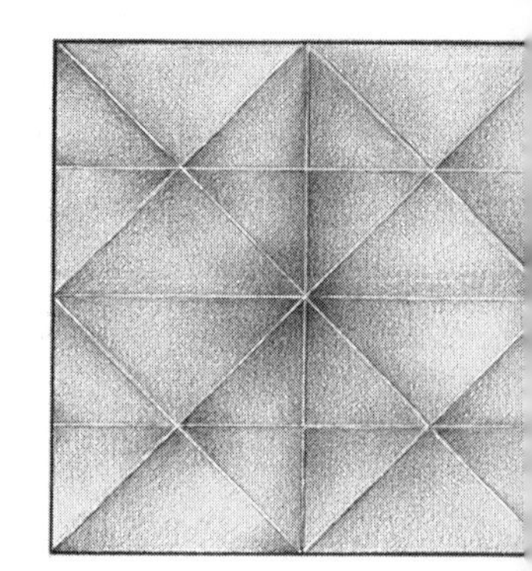

MATERIALS NEEDED FOR EACH STUDENT

At least eight squares of patty paper for each student
Origami journal
Scissors
Circles worksheet

GROUPING

Work with your group.

FOLDING A SQUARE INTO MANY PARTS

1. a. Find two different ways to fold a square into sixteen parts.

   b. Make one or more folds to create a polygon whose area is three-fourths of the area of the original square.

2. Fold a square paper so that when it's unfolded you have 64 equal-sized squares.

   a. What part of the area of the original square does one small square represent?

   b. Using fractions with a numerator of 1, list the fractional parts of the original square that can be represented by squares.

c. Using fractions with a numerator of 1, list the fractional parts of the original square that can be represented by rectangles (*not* rectangles that are also squares).

d. Show how you can use your folded square to find the sum of the sequence $\frac{1}{2} + \frac{1}{4} + \frac{1}{8} + \frac{1}{16} + \ldots$. (Hint: You may want to color in these fractional parts.)

e. Show how you can use your folded square to find the sum of the sequence $\frac{1}{4} + \frac{1}{16} + \frac{1}{64} + \ldots$. Explain your thinking. (Hint: Think about the squares and rectangles you listed in questions 2b and 2c.)

f. Show how you could use algebra to show the sums in questions 2d and 2e.

MORE SQUARE EXPLORATIONS

1. Try to come up with a folding sequence that results in a figure with an area $\frac{1}{3}$ the area of the original square. Do the same for $\frac{1}{5}$, $\frac{1}{6}$, $\frac{1}{7}$, and so on. Do you think it's possible to do any of these? All of these? Write a paragraph summarizing your findings.

2. If the area of your original square is four square units, what is the length of one side of the square that results when you fold down each corner? Use the Pythagorean theorem and your calculator.

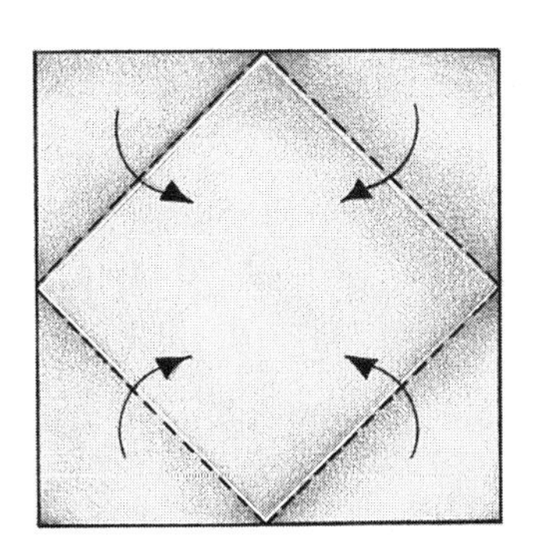

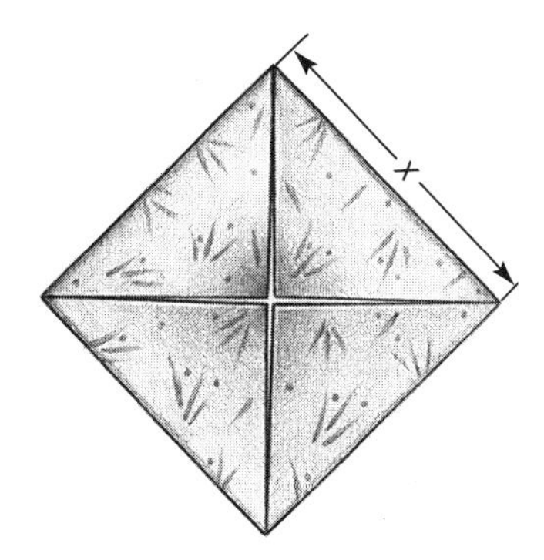

3. Using a square and the Pythagorean theorem, show how you can fold line segments whose lengths are irrational numbers like $\sqrt{2}$, $\sqrt{3}$, $\sqrt{5}$, and so on.

4. a. Cut out a circle. Fold it into a square. Explain how you know that it's exactly a square.

   b. Cut out another circle and fold it into a square a different way. What do you notice about the relationship between the two ways you found to fold a circular paper into a square? (Hint: Think about the blintz fold.)

5. In Japan, there is a special type of envelope folding called *tatō.* Follow the steps below to create a tatō.

   a. Cut out a paper circle.

   b. Fold and unfold the circle in half along a diameter.

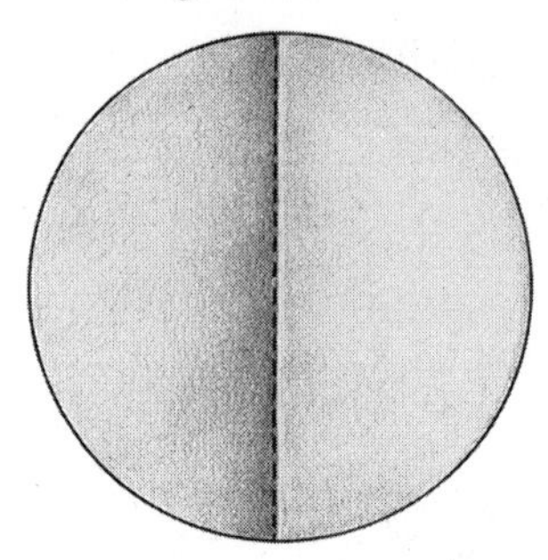

   c. Fold along another diameter perpendicular to the original diameter.

   d. Fold so that points *A* and *B* meet at the center of the circle.

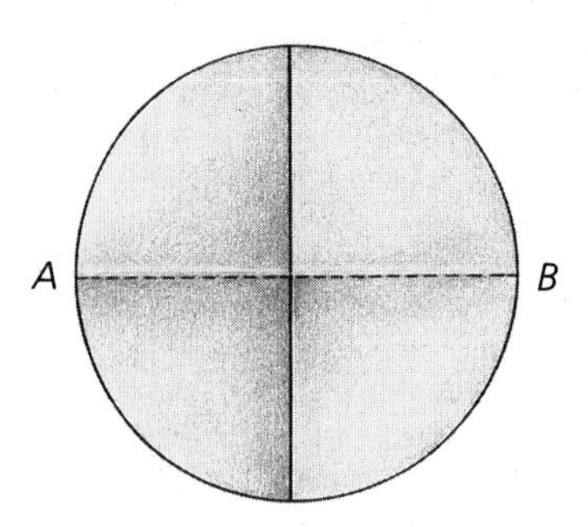

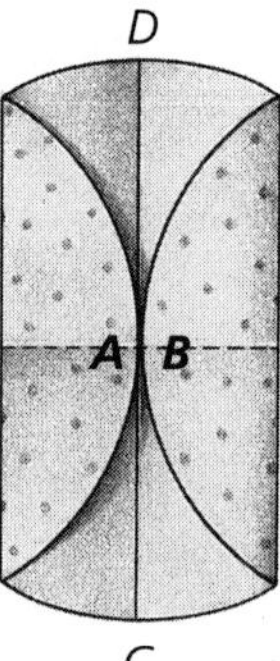

   e. Fold so that points *C* and *D* meet at the center.

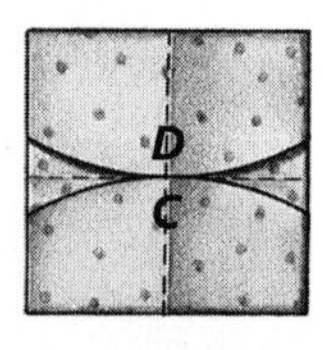

   f. Reverse the folds in the corner and tuck the right side of the lower flap under the left flap.

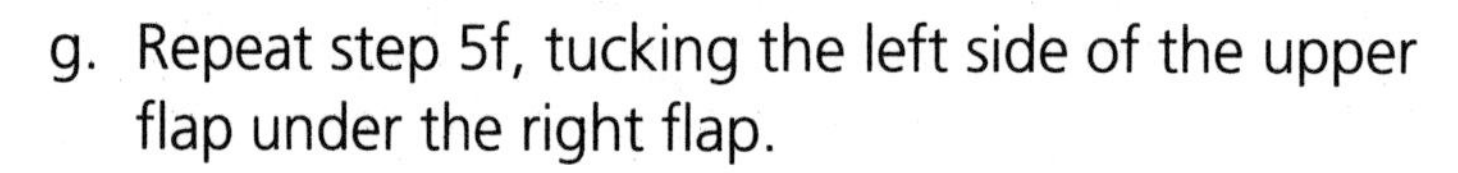

   g. Repeat step 5f, tucking the left side of the upper flap under the right flap.

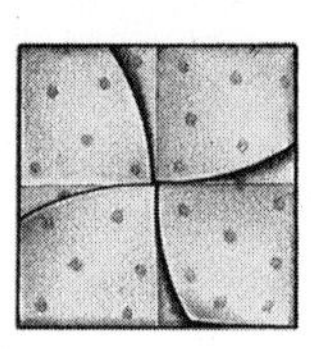

   h. You now have a Japanese tatō.

Folding a Square II

OBJECTIVES

To express the area of folded regions in terms of fractions
To add the terms of a sequence
To explore other fractional parts of a square
To discover some relationships between a circle and a square

MATERIALS NEEDED FOR EACH STUDENT

At least eight squares of patty paper
Student worksheet pages—one copy for each pair of students
Origami journal
Scissors
Circles worksheet—one sheet for each student

TEACHER MATERIALS

Overhead projector and waxed paper, or large paper for demonstration

TIME

One class period

GROUPING

Students should work in groups of four. Each student will do his or her own folding.

GENERAL INSTRUCTIONS

This is an optional activity. You may want to do only parts of this activity, or you may choose to let some students do it as an extra-credit assignment. Some of the questions are quite challenging and may involve mathematics that your students have not encountered yet. This activity is not a prerequisite for activities that follow.

ANSWERS

Folding a Square into Many Parts

1. a. Answers will vary. Two ways to fold a square into sixteen parts are shown below. There are other ways to do this.

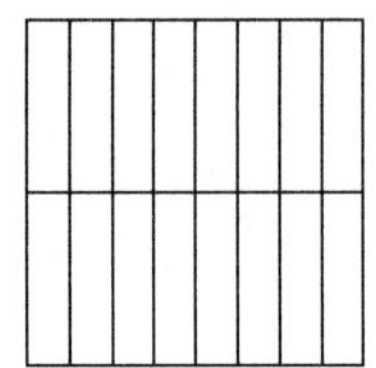

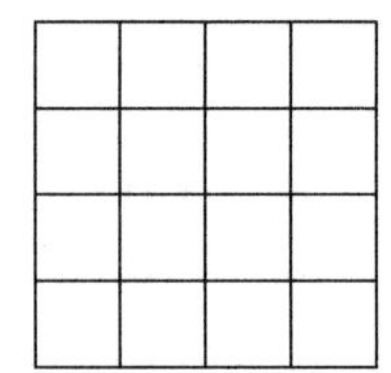

   b. Answers will vary. The following are two ways of folding an area that is three-fourths that of the original square.

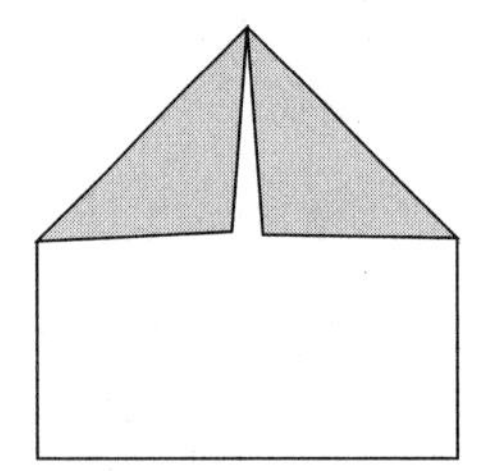

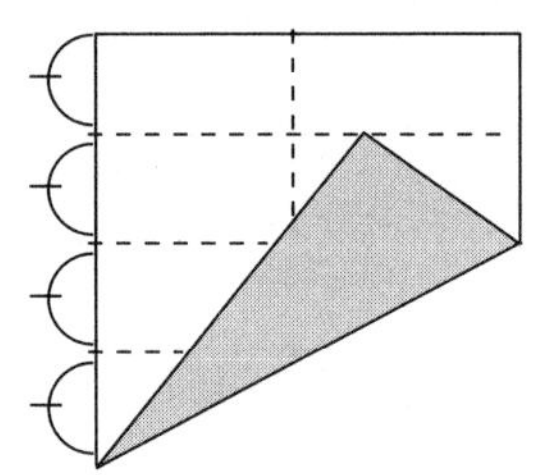

2. a. One small square represents $\frac{1}{64}$ of the area of the original square.

   b. The fractional parts represented by $\frac{1}{4}$, $\frac{1}{16}$, and $\frac{1}{64}$ are squares.

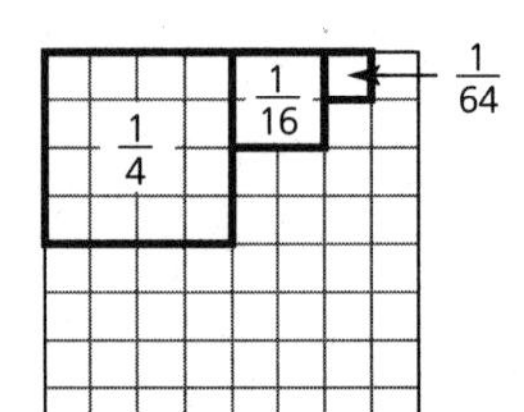

   c. The fractional parts represented by $\frac{1}{2}$, $\frac{1}{8}$, and $\frac{1}{32}$ are rectangles.

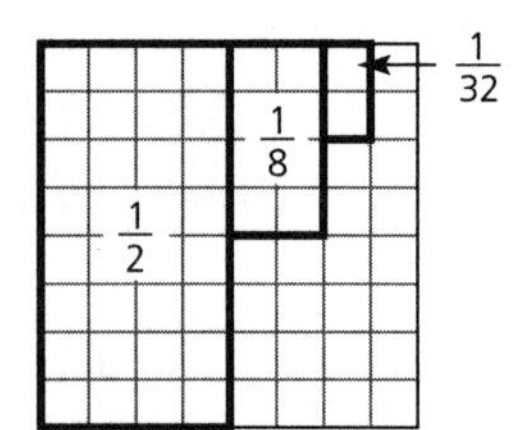

   d. If you color $\frac{1}{2}$, then $\frac{1}{4}$, then $\frac{1}{8}$, and so on, you can see that the sum of this sequence approaches 1.

   The sum of the series $\frac{1}{2} + \frac{1}{4} + \frac{1}{8} + \ldots$ can also be found using the formula $S = \frac{a}{1-r}$, where a is the first term and r is the common ratio.

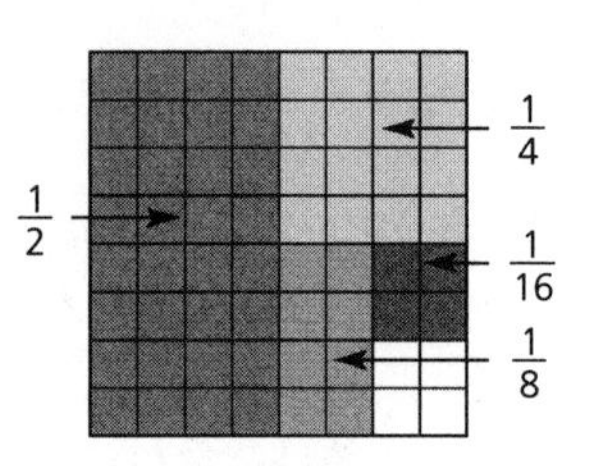

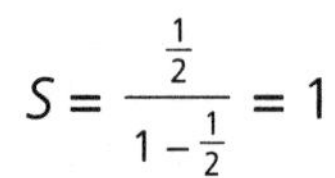

$$S = \frac{\frac{1}{2}}{1 - \frac{1}{2}} = 1$$

e. The sum of the sequence is $\frac{1}{3}$. One intuitive geometric explanation is that the area of each rectangle is double the area of a square. That is, the area of the rectangle that represents $\frac{1}{2}$ is double the area of the square that represents $\frac{1}{4}$. The area of the rectangle that represents $\frac{1}{8}$ is double the area of the square that represents $\frac{1}{16}$. It follows that the area of all the rectangles is twice the area of all the squares. Therefore, if the area of the large square is 1, then the area of all the rectangles is $\frac{2}{3}$ the total area. In the same way, the area of all the squares is $\frac{1}{3}$ the total area. Therefore, the sum of the areas of all the squares is $\frac{1}{3}$.

f. The solution to question 2d includes an algebraic method for finding the sum. In like manner, the sum of the series $\frac{1}{4} + \frac{1}{16} + \frac{1}{64} + \ldots$ can be found using the formula $S = \frac{a}{1-r}$, where a is the first term and r is the common ratio.

$$S = \frac{\frac{1}{4}}{1 - \frac{1}{4}} = \frac{1}{3}$$

More Square Explorations

1. Answers will vary.
2. $\sqrt{2}$ units
3. Answers will vary.
4. There are two ways to fold a square when you start with a circular paper.

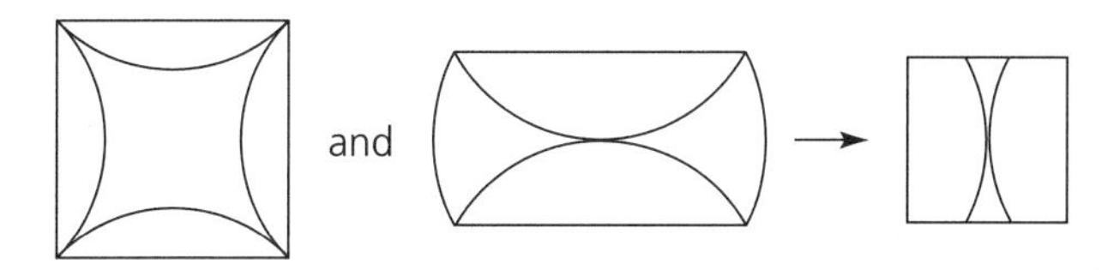

The relationship between the two different results is shown in the diagram at the right. The area of the smaller square is half the area of the larger square. When the smaller square is placed on top of the larger one, they look like a blintz fold before the folding has taken place.

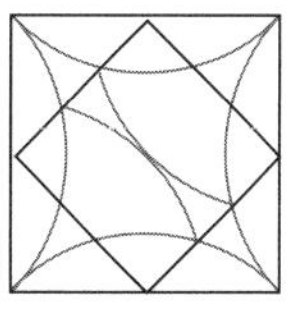

5. Students fold a tatō.

Circles Worksheet

My Experiences with Origami

Vernon Isaac

ernon Isaac

Vernon Isaac is a computer specialist at General Electric who hails from Brooklyn, New York. He's been interested in origami for about three and a half years, and from time to time he teaches origami classes. He likes money folds ecause if he gets tired of his hobby, he can pend his models! Vernon's other hobbies nclude chess, horticulture, umpiring, and icycling.

ly interest in origami started one day when a iend of mine showed me a paper bow tie he ad made. I was unable to duplicate the bow tie ut was still interested in the idea of folding a nodel. I modified my friend's bow tie and came p with a dollar-bill bow. My daughter, Yolanda, vho was then nine, had been given some rigami books by her cousin, so I asked her to how me the books and the results of some of er models. Seeing Yolanda's books and models vas enough to whet my folding appetite, and ve went to the library to get more material on he subject of origami. (Thank goodness for braries! If you can read—a skill that will give ou great enjoyment throughout your life—and an get to a library, you can find out about any opic under the sun.) We tried several models rom the books but with little success. Not being ne to give up, I read each book from cover to over. The address of the Origami Society at the Auseum of Natural History was in the back of one of the books, and one morning when Yolanda and I were in need of a new adventure, we went to the museum and found the society.

My vision was to make a rose out of dollar bills. At the society I met a folder named Myer Gotz, and he told me that although he'd never seen a dollar-bill rose, he knew other dollar folds. Myer showed me a dollar elephant and two other designs, then sent me home to practice. I worked on my new toy and finally was able to make the dollar elephant, but with one catch—the elephant kept falling over face first. I tried folding the model many times and always came up with the same result. To compensate for this balance problem and to cover my mistake, I put another head on the elephant, facing in the opposite direction, and called my model a Siamese elephant. Nobody had ever seen a Siamese elephant before—of course not: it was my creation!—and I was off and running. (I have since mastered the dollar elephant, and I've deemed my Siamese elephant the meanest animal in the world!)

I was still interested in learning how to fold a dollar-bill rose, and I found out how to do it at the 1995 Origami Convention. At the convention, I attended several money folding classes and during the last class of the final day, I met a woman named Carol from upstate New York. I told her of my plight: all I wanted to do was make a dollar-bill rose. She said, "Is that all?" and sent me to get some wire and pliers. She then promptly showed me how to make a rose. I went home and practiced, making altogether about twenty dollar-roses.

This is my vision for my roses: *She wore two*

Siamese elephant designed and folded by Vernon Isaac.

bows in her hair (my bows), and she danced around the room with castanets in each hand and a beautiful dollar rose held tightly between her teeth. Her name was Sunshine, and she was exquisite. The dollar rose is now my masterpiece and duly named Sunny. Every year at the Origami Convention, I design a variation of my rose. My most recent dollar rose is now on display at the Origami Society in the Museum of Natural History and is called *The Midnight Rose,* or *Sunny Under Glass.* It is one and a half feet tall, made of two-dollar bills, and enclosed in a glass bell dome.

Since I designed my first models, I've broadened my horizons to design a set of nested boxes out of business cards. To include my friends who don't have business cards, I use floral, colored cards. My nested-box design features a set of nine boxes that nest (fit) into one another. The ninth and smallest box measures just under one-quarter of an inch. One day I am going to make a tenth box that measures about one-sixteenth of an inch—this box will be the final level. Other business-card designs include the cube octahedron, which is composed of twenty-four sides and twenty-four angles. I've made the cube octahedron in many sizes, including one that was extra large and one to fit into each of the nine nested boxes!

To me origami is a free form of expression with paper. I'm also interested in origami's symmetry of sides and angles, its mathematics (including its geometry), and its basic concept that if you fold this way and that way, the final result is a complete model made of paper. The beauty of it all—AH!!!

I would like to thank the many wonderful people I have met through origami; my wife Karen for her patience in watching me fold; and my children, especially Yolanda, who is my biggest fan and best critic—she tells me exactly what she sees.

ACTIVITY 4

Ori Money in Two Dimensions

Although the square is the most popular shape for origami paper, some origami folders use circular paper, and many enjoy folding currency. They fold many interesting geometric shapes as well as very intricate animal models. David Masunaga is an origami folder and designer who likes to fold money. He even creates three-dimensional models using dollar bills. In this activity you will transform a dollar bill into an equilateral triangle using ideas written about by Jean Pedersen in a letter to the editor of *Mathematics Magazine* 61, no. 4 (October 1988): 270. Even though a dollar bill is rectangular and not square, when you fold one it is still considered origami.

MATERIALS NEEDED FOR EACH STUDENT

A one-dollar bill
Origami journal

GROUPING

Work with a partner. Each person should do his or her own folding.

FOLDING INSTRUCTIONS AND QUESTIONS

When you are folding, think about the answers to the folding questions asked at each step. When you have finished folding, answer these questions in your origami journal.

1 Fold the dollar bill in half lengthwise and unfold.

2 Fold the top left-hand corner down so that point A is on the crease. The fold must also go through point D.

Which angle measure do you know for sure in the folded triangle? How many degrees are in this angle?

3 Use one edge of the triangle you just made as your new fold line.

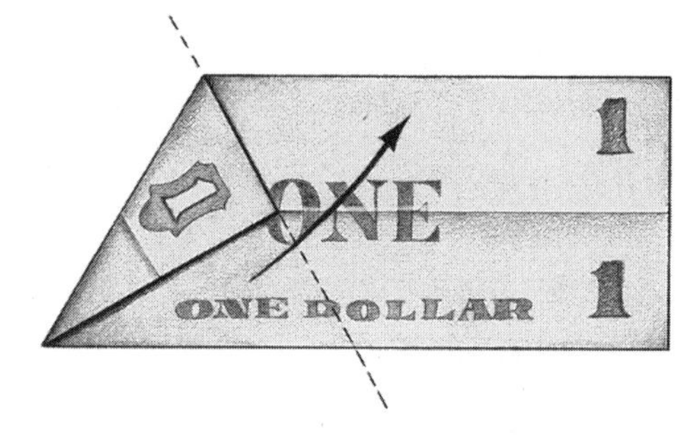

What kind of triangle does this new triangle appear to be?

4 Use the edge of the triangle as your next fold line.

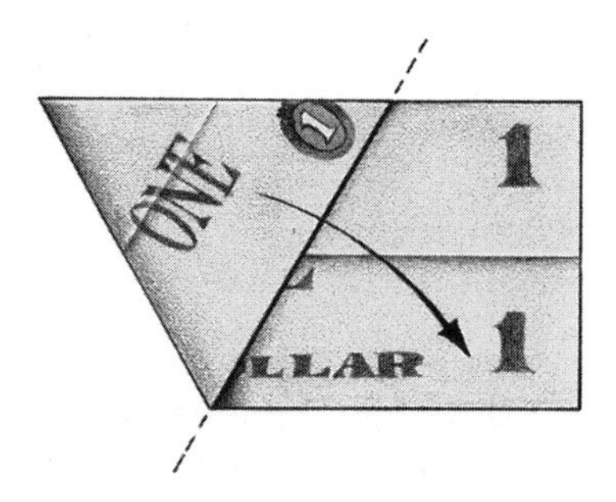

5 Use the edge of the new triangle as your next fold line.

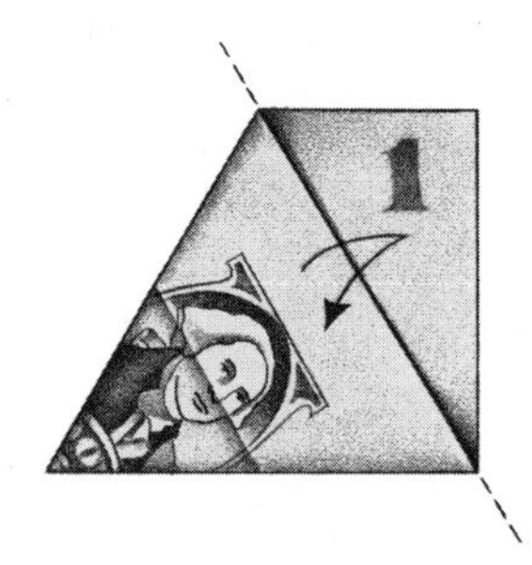

How much of the area of the larger triangle does the last flap cover?

6 Unfold the last fold. Tuck the flap into the pocket.

7 The completed piece looks like this.

Remember to answer each of the folding questions in your origami journal.

EXPLORING YOUR MODEL

Record your answers to the exploring questions in your origami journal.

1. How can you show that the triangles you have been folding are equilateral triangles? (Hint: Use what you know about the 30-60-90 triangle you folded in the first step.)

2. Unfold your dollar bill. Use what you know about parallel lines, alternate interior angles, and vertical angles to find the measure of each labeled angle in the following diagram. Make a sketch of the diagram in your origami journal and label the measure of each angle.

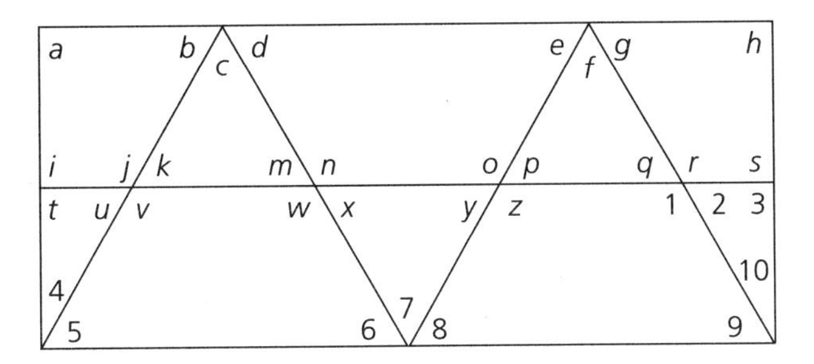

3. When you unfold your dollar bill, how many equilateral triangles do you see?

4. Try the ori money folding sequence with a strip of adding machine tape or a half sheet of paper cut the long way. Do you always get equilateral triangles? Do you always get the same number of equilateral triangles? Explain why or why not.

5. Do you think that you would get the same number of equilateral triangles with currency from another country? Explain why or why not.

Ori Money in Two Dimensions

OBJECTIVES

To fold angles on a rectangle
To investigate the 30-60-90 triangle
To investigate the equilateral triangle
To use parallel lines, alternate interior angles, and vertical angles to find the measure of an angle
To learn that any type of paper, including rectangular paper, can be used for origami

MATERIALS NEEDED FOR EACH STUDENT

A one-dollar bill
Student worksheet pages—one copy for each pair of students
Origami journal

TEACHER MATERIALS

A one-dollar bill, or a piece of waxed paper the same size as a dollar bill
Overhead projector

TIME

One class period

GROUPING

Students should work with a partner. Each person will complete his or her own ori money folds.

GENERAL INSTRUCTIONS

If students have difficulty following the folding instructions, you can demonstrate the folding sequence using waxed paper on an overhead projector or a dollar bill. Encourage students to think about the questions asked in the folding instructions while completing the folding sequence. After students have completed the folding, remind them to go back and answer the questions in their origami journal.

ANSWERS

Folding Questions

1. [No question.]
2. The right angle. Its measure is 90°.
3. An equilateral triangle.
4. [No question.]
5. The flap covers half the area of the equilateral triangle.

6–7. [No questions.]

Exploring Your Model

1. Students should notice that the first triangle they folded is a 30-60-90 triangle. By folding, they can show that the area of this right triangle is half the area of each equilateral triangle.
2. Angles *a, h, i, s, t,* and 3 all measure 90°.

   Angles *b, c, d, e, f, g, k, m, p, q, u, x, y,* 2, 5, 6, 7, 8, and 9 all measure 60°.

   Angles *j, n, o, r, v, w, z,* and 1 all measure 120°.

   Angles 4 and 10 measure 30°.
3. There are six equilateral triangles: three large and three small.
4. There will always be equilateral triangles. The number of equilateral triangles depends on the ratio of the length and width of the paper strip.
5. The ratio of the length and width of different currencies varies, so the number of triangles will also vary.

ACTIVITY 5

Triangular Measuring Tool

Kunihiko Kasahara, who has written many books on origami, has shown that with four folds you can make a useful tool for measuring eight different angles. If you ever forget your protractor, you will still have a lot of measuring power if you can find a square piece of paper. The folding process for making the triangular measuring tool is easy if you take it step by step.

MATERIALS NEEDED FOR EACH STUDENT

One square of origami or patty paper
Origami journal

GROUPING

Work with a partner. Each person should fold his or her own measuring tool.

FOLDING INSTRUCTIONS AND QUESTIONS

When you are folding, think about the answers to the folding questions asked at each step. When you have finished folding, answer these questions in your origami journal.

1 Fold the paper in half and unfold.

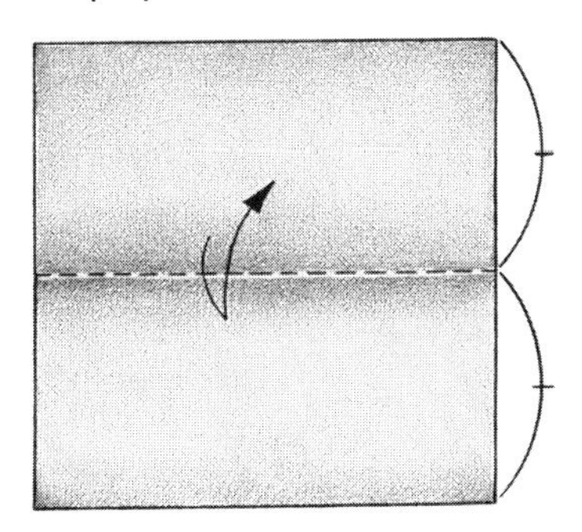

a. *What do the single hatch marks on the arcs mean?*

b. *How do the length and width of each rectangle compare to the side length of the square?*

2 Fold the top right corner down so that vertex *A* falls on midsegment *BC*. Be sure the fold passes through vertex *D*.

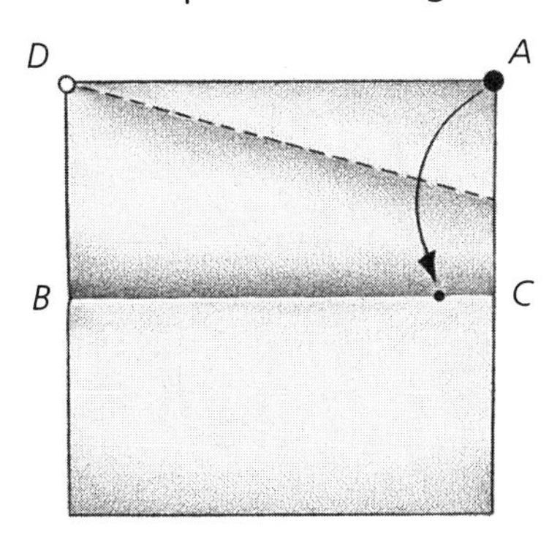

What kind of triangle did you just fold?

3 Fold the bottom left corner up so that it meets the top right corner of the square.

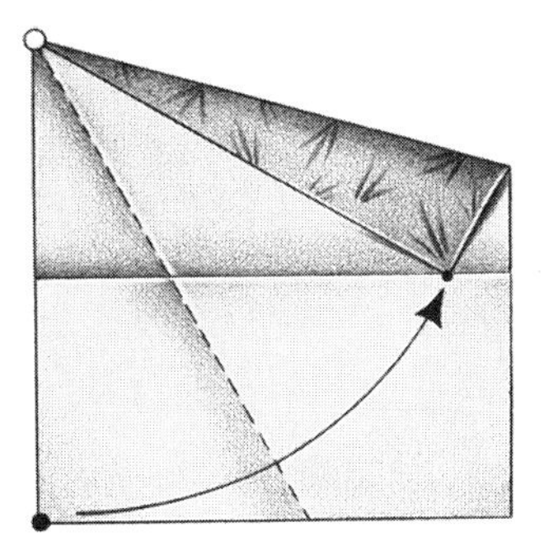

What kind of triangle is formed by the dashed fold line pictured above?

4 Fold up the bottom triangle.

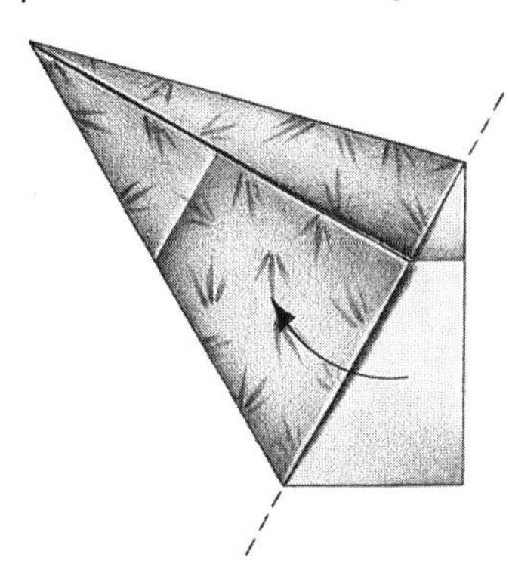

What do all of the triangles in the picture above have in common?

5 You have folded a triangular measuring tool.

Remember to answer each of the folding questions in your origami journal.

EXPLORING YOUR MODEL

Record your answers to the exploring questions in your origami journal.

1. Unfold your angle measuring tool and find the measure of each angle formed by the fold lines. Write these angle measures on your angle measuring tool, and keep the tool in your origami journal for future reference. Explain how you found each angle measure.

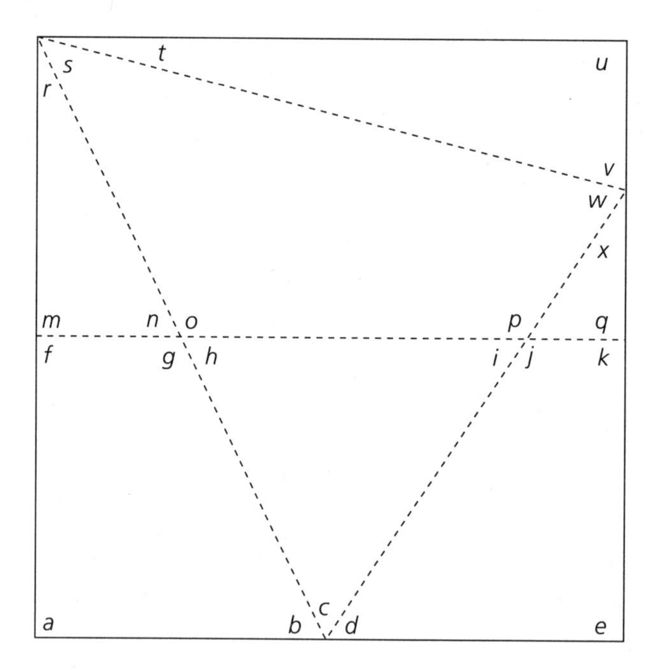

2. Make a list of the different angle measures you found on your measuring tool.

3. Architects call a 30-60-90 triangle a 30° triangle and a 45-45-90 triangle a 45° triangle. Explain why you think they do this.

4. Use your triangular measuring tool to measure the interior angles and the exterior angle in each polygon that follows. To measure some of the angles, you may need to use a combination of two angles on your tool.

Regular polygon	Measure of each interior angle	Measure of the exterior angle
Equilateral triangle		
Regular hexagon		
Regular octagon		
Regular dodecagon		

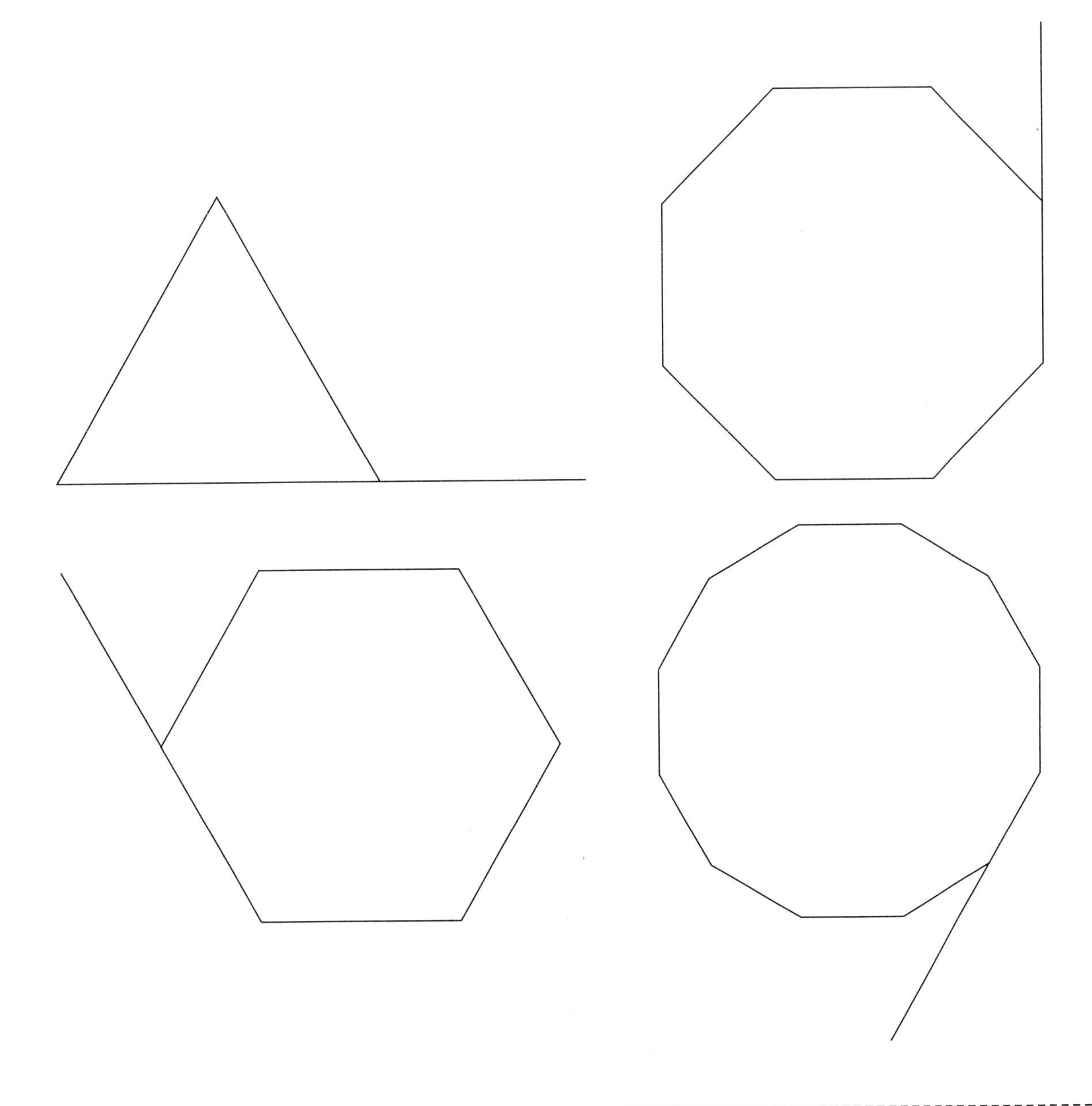

Triangular Measuring Tool

OBJECTIVES

To explore relationships between angle measures
To explore different types of right triangles
To apply the sum of the angles of a triangle theorem
To fold a triangular measuring tool
To appreciate the power, simplicity, and economy of origami

MATERIALS NEEDED FOR EACH STUDENT

One square of origami or patty paper
Student worksheet pages—one set for each pair of students
Origami journal

TEACHER MATERIALS

A square piece of waxed paper or a large paper square
Overhead projector

TIME

30 minutes

GROUPING

Students should work with partners. Each student will fold his or her own measuring tool.

GENERAL INSTRUCTIONS

Patty paper works well for this activity because students can write the angle measures directly on the folded paper and keep the patty paper in their origami journal for future reference.

If students have difficulty following the folding instructions, you can demonstrate the folding sequence using waxed paper on an overhead projector or using a large paper square. Encourage students to think about the questions asked in the folding instructions while completing the sequence. After students have completed the folding, remind them to go back and answer the questions in their origami journal.

ANSWERS

Folding Questions

1. a. The hatch marks on the arcs indicate congruent lengths.

   b. The long side of the rectangle and the side of the square are equal in length. The width of the rectangle is half the side length of the square.

2. A right scalene triangle.
3. A right scalene triangle, 30-60-90 triangle.
4. They are all right scalene triangles.
5. [No question.]

Exploring Your Model

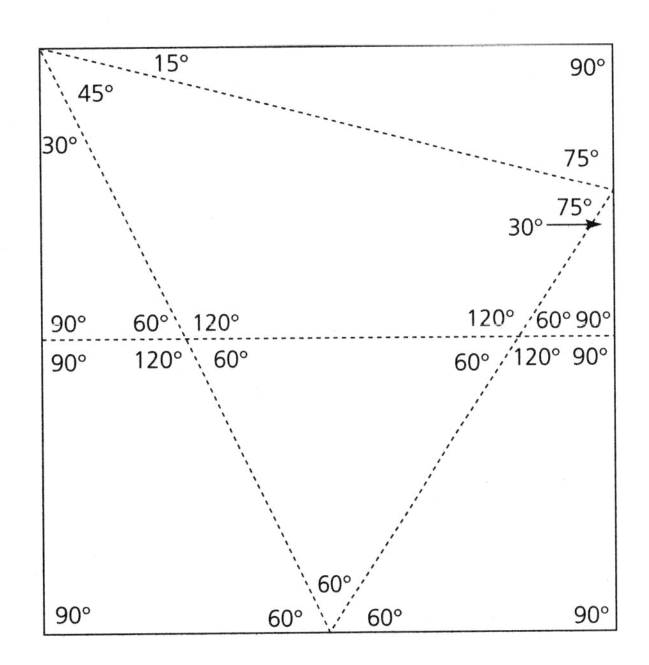

1. Students can record the angle measures directly on the diagram. They should record their rationale for each angle measure in their origami journal. Answers will vary for the rationales for the angle measures. Encourage students to share their reasoning.
2. The different angles on the measuring tool are 15°, 30°, 45°, 60°, 75°, 90°, 105°, 120°, 150°, and 180°.
3. Architects use right-triangle tools. If you know the measure of one of the acute angles of a right triangle, then you know the measure of the other acute angle.
4.

Regular polygon	Measure of each interior angle	Measure of the exterior angle
Equilateral triangle	60°	120°
Regular hexagon	120°	60°
Regular octagon	135°	45°
Regular dodecagon	150°	30°

ACTIVITY 6

Folding a Regular Hexagon

The hexagon shape occurs often in nature. Perhaps you have seen pictures of snowflakes or honeycombs that show hexagons. In traditional geometry, you can construct a regular hexagon using a compass, straightedge, and pencil. In the world of origami, you can fold a square into a regular hexagon. The folding sequence in this lesson was diagrammed by Kunihiko Kasahara in his and Toshie Takahama's *Origami for the Connoisseur.*

MATERIALS NEEDED FOR EACH STUDENT

One square of origami or patty paper
Origami journal

GROUPING

Work with a partner. Each person should fold his or her own hexagon.

FOLDING INSTRUCTIONS AND QUESTIONS

When you are folding, think about the answers to the folding questions asked at each step. When you have finished folding, answer these questions in your origami journal.

1 Fold the square in half on the diagonal.

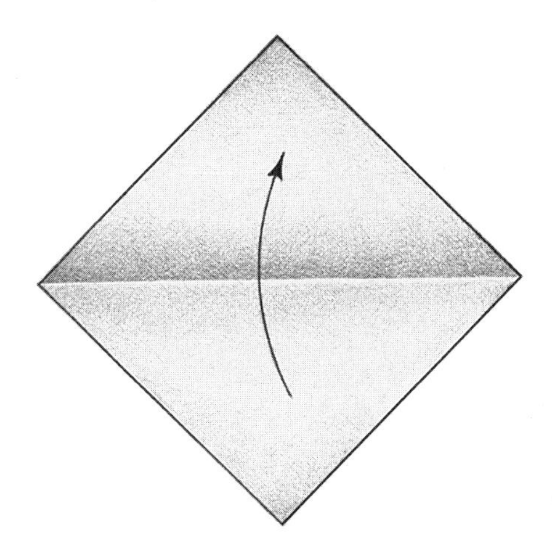

Two congruent triangles form a square. What kinds of triangles are they?

2 Rotate the triangle so that the fold is horizontal. Make a short crease to mark the midpoint of side *BC*.

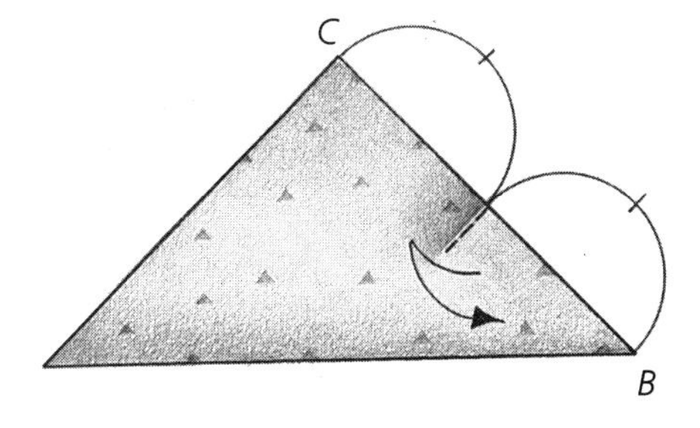

Which part of the picture tells you to bisect the leg of the triangle?

3 Fold and unfold the perpendicular bisector of segment *MC*.

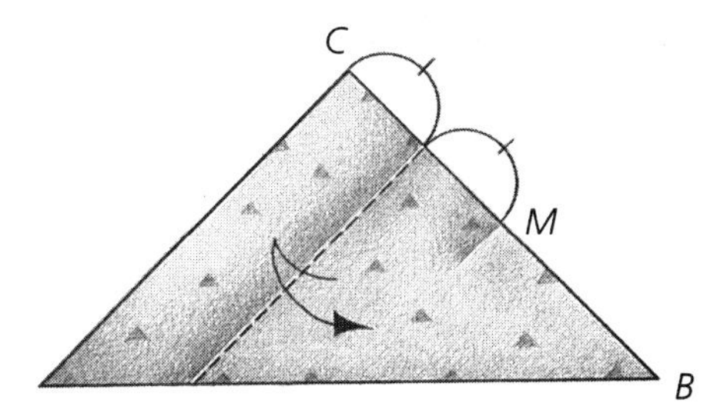

4 Fold and unfold the perpendicular bisector of segment *AB*.

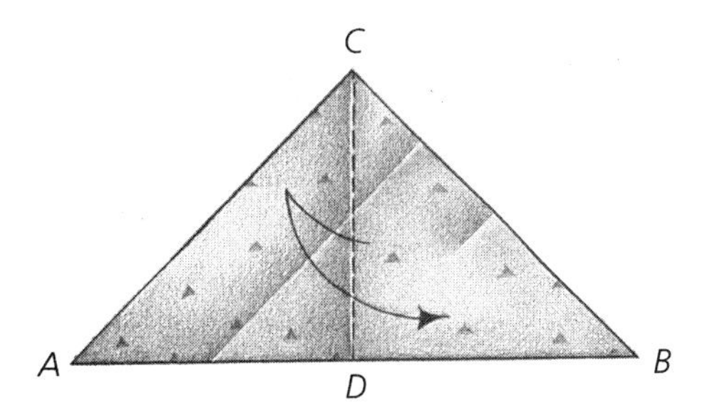

a. *How do you know ΔABC is a right isosceles triangle? What are the measures of its angles?*

b. *Name two triangles that are similar to ΔABC.*

5 Fold so that point *M* lies on segment *EF*. The fold should pivot on point *D*.

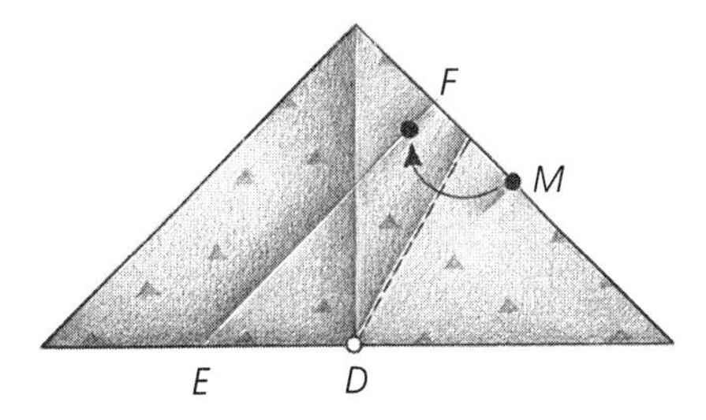

6 Fold the flap under along the line defined by the side of the triangle you just folded.

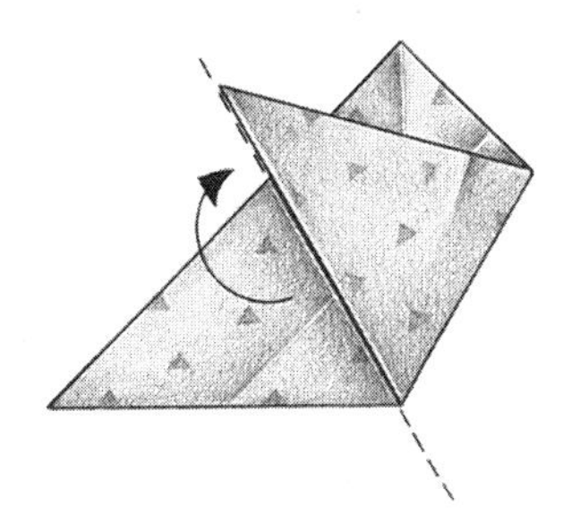

7 Fold down and unfold to form a crease.

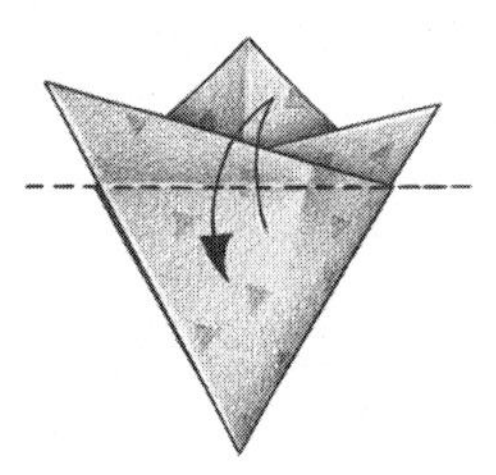

8 Unfold completely and find the hexagon.

Remember to answer each of the folding questions in your origami journal.

EXPLORING YOUR MODEL

Record your answers to the exploring questions in your origami journal.

1. Use a pencil or marker to outline the hexagon on your unfolded paper. Use the triangular measuring tool you made in Activity 3, or what you know about geometry, to find the measure of each interior angle in the hexagon.
2. Use your triangular measuring tool and what you know about geometry to find the measures of as many angles as you can on your unfolded hexagon paper. Label the angle measures on your paper.
3. If your folds were accurate, you folded a regular hexagon. Which step of the folding instructions most clearly shows you that your hexagon is regular? Explain.
4. What other regular polygon do you see in the folds of your hexagon?
5. The folds on your hexagon show some lines of reflection symmetry. Which lines are missing? Fold them in. How many lines of reflection symmetry are there altogether? Describe them.
6. Describe your hexagon in terms of rotational symmetry.

EXPLORING FURTHER

In Japan, the art of cutting rotationally symmetric designs in paper is called *kirigami.* In this exploration you will use what you have learned so far in this activity to create a snowflake design.

1. Follow the directions for folding a regular hexagon through step 7, and cut along the last fold you made. Then make a new fold perpendicular to the cut side. To cut out the design, start cutting on the "spine side," but be sure to leave some part of the spine intact. After making the first cut on the spine side, you can make a cut on the nonspine side, or you can continue cutting on the spine side. You can even cut off the tip if you leave part of each side intact. Unfold completely to reveal your snowflake design.

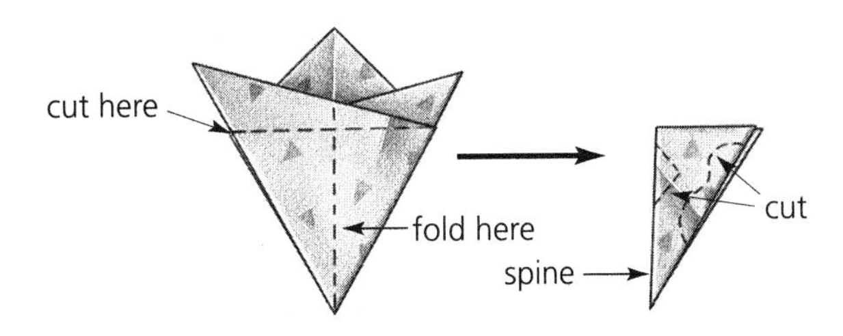

2. When you unfold your creation, you should have a snowflake. What kind of rotational symmetry does your snowflake have? Explain why the extra fold was necessary.

Folding a Regular Hexagon

OBJECTIVES

To fold a regular hexagon
To fold 30° and 60° angles
To recognize the relationship between the equilateral triangle and the regular hexagon
To find the measure of the interior angle of a regular hexagon
To explore the reflection and rotational symmetries of a regular hexagon

MATERIALS NEEDED FOR EACH STUDENT

One square of origami or patty paper
Student worksheet pages—one copy for each pair of students
Origami journal

TEACHER MATERIALS

Overhead projector and waxed paper, or large square paper for demonstration

TIME

30 minutes

GROUPING

Students should work with partners. Each student will fold his or her own hexagon.

GENERAL INSTRUCTIONS

Encourage students to help each other with the folding instructions. This folding activity can be done with patty paper. If students didn't do Activity 2, before asking them to answer questions 5 and 6 under Exploring Your Model, be sure to discuss reflection and rotational symmetry.

ANSWERS

Folding Questions

1. Each of the triangles formed is an isosceles right triangle.
2. The two congruent arcs tell you to bisect the leg of the triangle.
3. [No question.]
4. a. You know that ΔABC is a right isosceles triangle because segment *CD* is perpendicular to segment *AB,* and segment *AC* is congruent to segment *BC.* The measures of its angles are 45°, 45°, and 90°.

   b. ΔACD and ΔBCD.

5–8. [No questions.]

Exploring Your Model

1. Each interior angle of a regular hexagon has a measure of 120°.
2. See figure.
3. Step 7 of the folding process shows in a purely visual way that the hexagon is regular. In this step, all six congruent triangles making up the hexagon are superimposed on each other. From this step, students can see that the sides of the hexagon are congruent.
4. The equilateral triangle can be added to the list of regular polygons.

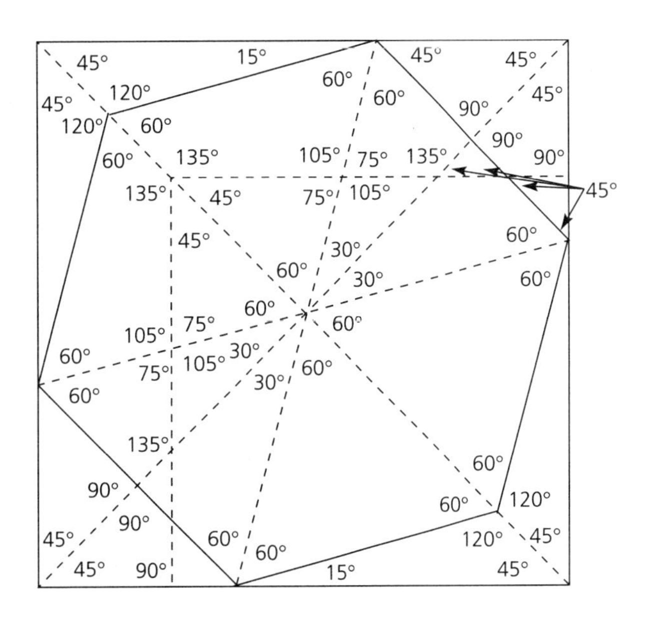

5. A hexagon has six lines of reflection symmetry. Three of the lines of symmetry connect pairs of vertices. The other three symmetry lines are perpendicular bisectors of the sides. Two of these bisectors are missing in the folds.
6. The hexagon has 6-fold rotational symmetry.

Exploring Further

1. Students create a "snowflake."
2. The snowflake has 6-fold rotational symmetry. If you didn't make the extra fold, the snowflake would only have 3-fold rotational symmetry.

ACTIVITY 7

Free Folding an Equilateral Triangle

Many origami folders enjoy following diagrams and folding models created by other origami designers. Some folders are more interested in creating new designs themselves. Some designers develop their new creations purely by intuition, while others, relying on an understanding of geometry, use their knowledge of angle relationships. In this activity, you will make up your own steps for folding an equilateral triangle. You folded an equilateral triangle before while making a hexagon. However, this time your final result will be the largest equilateral triangle that will fit on your square piece of paper. (Your experience folding a dollar bill may also come in handy here.) Before you start, think about the properties of an equilateral triangle.

MATERIALS NEEDED FOR EACH STUDENT

One to three squares of origami or patty paper
Origami journal

GROUPING

Work by yourself or with a partner.

FOLDING AN EQUILATERAL TRIANGLE

Fold a square sheet of paper so that your folds show the largest equilateral triangle possible. Record your answers in your origami journal.

1. How long did you work on your folds? How difficult did you find this task?

2. Write out your instructions for folding an equilateral triangle, and draw sketches to go with your instructions.

3. Write about how you know your triangle is equilateral.

4. Have another person try your instructions. Revise them if necessary.

EXPLORING YOUR MODEL

Record your answers to the exploring questions in your origami journal.

1. Use your folded equilateral triangle to explore its symmetry.

   a. How many lines of symmetry does the equilateral triangle have? Describe these lines of symmetry.

   b. Describe the rotational symmetry of the equilateral triangle.

2. Draw a diagram showing how you can divide your large equilateral triangle into four smaller equilateral triangles by making three more folds.

3. Draw a sketch of your unfolded square. Find the measure of each angle determined by the folds, and write the measures on your sketch.

Free Folding an Equilateral Triangle

OBJECTIVES

To fold an equilateral triangle
To create and write directions for an original series of folds
To see the relationship between the square and the equilateral triangle

MATERIALS NEEDED FOR EACH STUDENT

One to three squares of origami or patty paper
Student worksheet page—one for each pair of students
Origami journal

TEACHER MATERIALS

None

TIME

One class period or less

GROUPING

Students can work individually or with a partner.

GENERAL INSTRUCTIONS

This activity will provide students with an opportunity to do some problem solving as they try to devise a method for using a square paper to fold an equilateral triangle. Encourage students to apply what they have learned in previous lessons about angles and folding.

ANSWERS

Folding an Equilateral Triangle

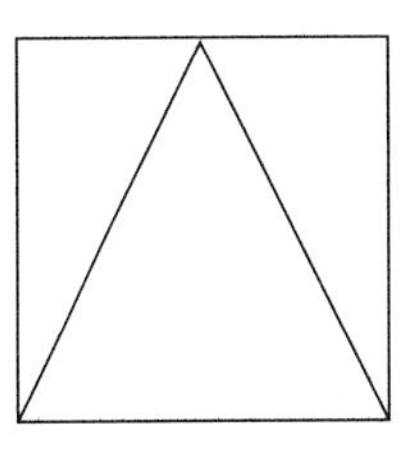

1– 4. Answers will vary.

Student folding sequences will vary. Be sure to allow time for students to share their results. Students often think that the triangle shown at the right is equilateral, but of course it isn't. This activity will help them see more clearly the relationship between the square and the equilateral triangle.

The folding sequence on page 42 shows one way to fold the equilateral triangle.

Exploring Your Model

1. a. The equilateral triangle has three lines of reflection symmetry. Each line of symmetry is both an angle bisector and the perpendicular bisector of a side.

   b. The equilateral triangle has 3-fold, or 120°, rotational symmetry.

2. The diagram below shows the three folds.

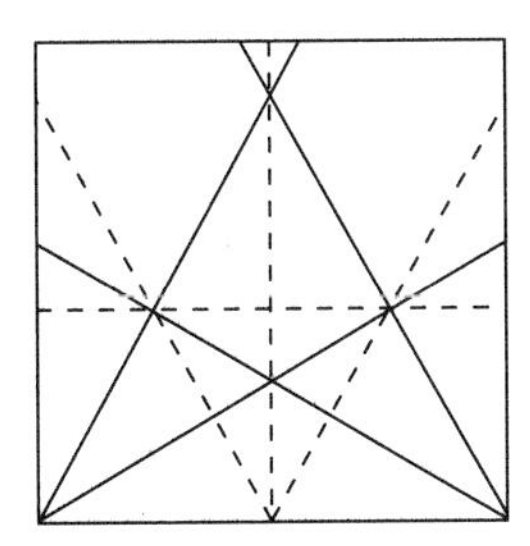

3.

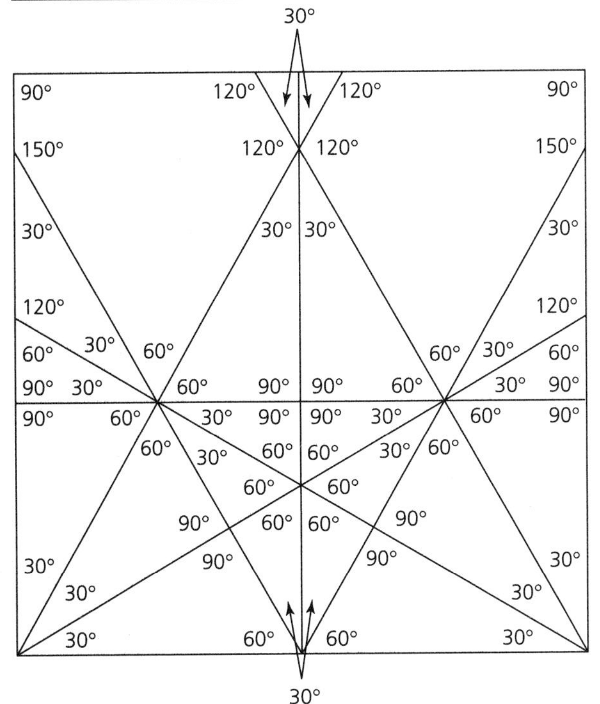

ONE WAY TO FOLD AN EQUILATERAL TRIANGLE

1. Fold and unfold.

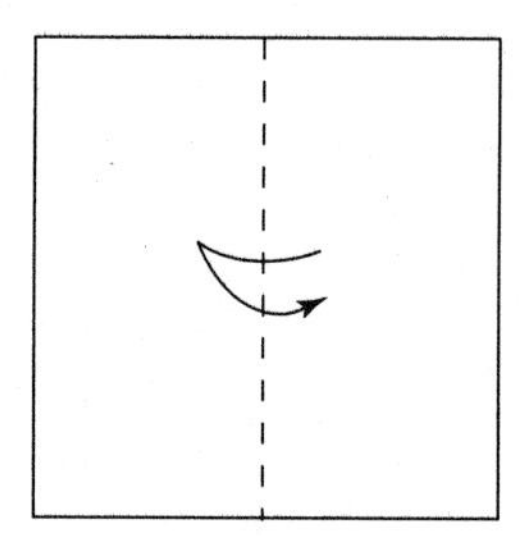

2. Fold so the dots meet. This fold shows you how long the side must be in order to be equal in length to the base.

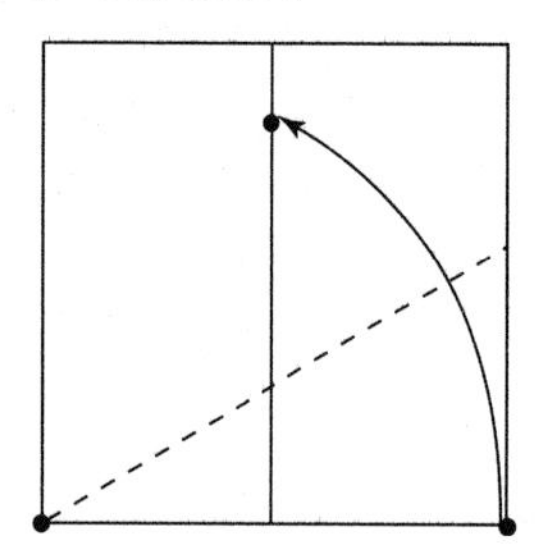

3. Fold and unfold.

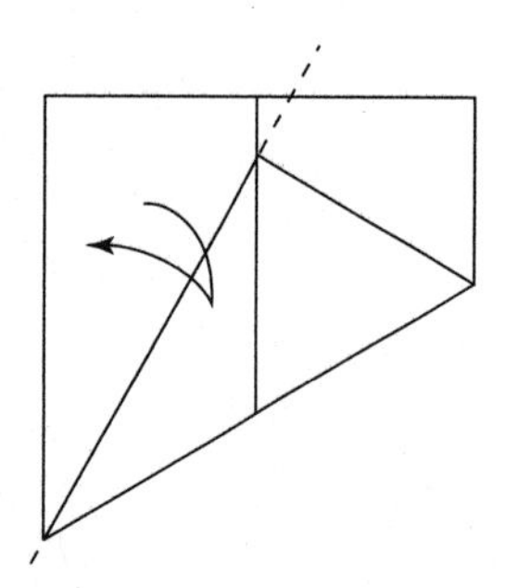

4. Open.

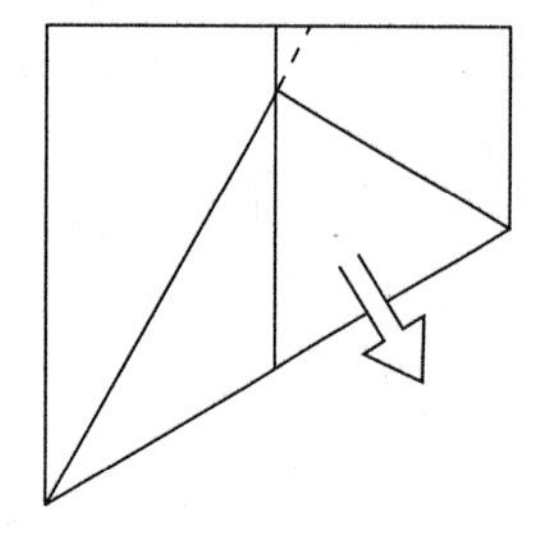

5. Fold so the dots meet.

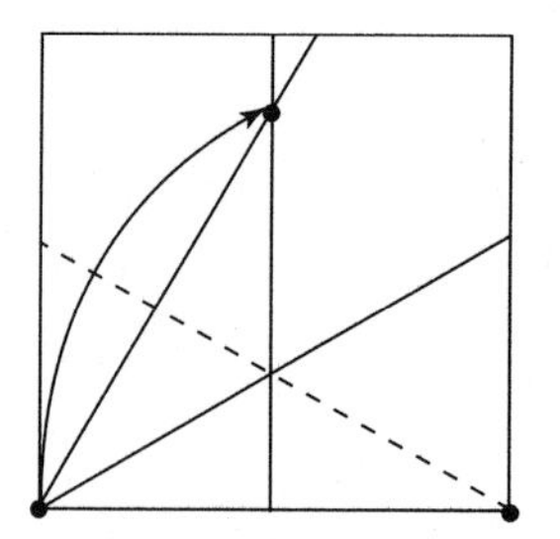

6. Fold and unfold.

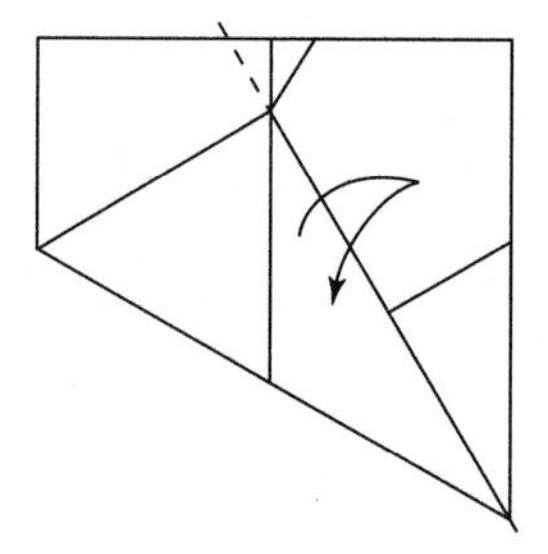

7. Open.

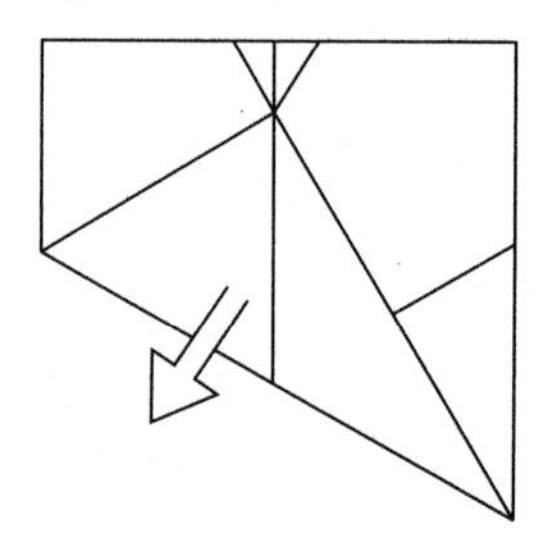

8. You have an equilateral triangle.

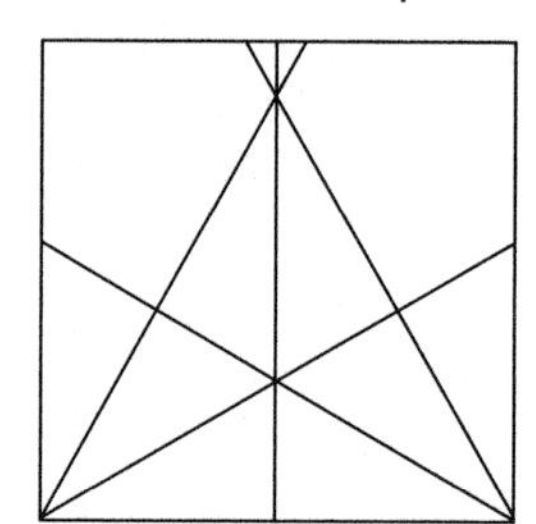

My Experiences with Origami

Jeremy Shafer

ıy Shafer

Jeremy Shafer is a well-respected origami designer from Berkeley, California, and editor of the Bay Area Rapid Folders *newsletter. His other interests include juggling, unicycling, clowning, and dance. The following about Jeremy's* ɾ*iences with origami is condensed from* ʼs *he wrote for his upcoming book,* Origami ıe 21st Century. *The complete essays are* ɜ*d on his website, located at* ɣ*.krmusic.com/barfup/html/jeremy.htm.*

ɪmost folders throughout the world, I was ɖuced to origami on the playground. Paper ɜnes, cootie catchers, and water bombs are ɟround folds I learned in elementary school. n I was ten, my parents gave me my first ɪnced origami book, *Origami for the* *usiast* by John Montroll. The first model I from the book was the grasshopper. ɔugh my finished model did not look nearly ɔod as the model pictured in the book, it my first exposure to advanced folding ɪiques and it allowed me to take off into a world of designing origami. It was from ʼolds of a grasshopper that I designed my ɒorigami model, a five-headed bird. I looked e grasshopper base and noticed that it ɐd just like a bird base but with four extra ɪts. I thought, why not turn these extra ts into heads?

From playground folds and the few Montroll models I had done, I learned enough folding techniques to explore origami on my own. I simply liked folding by instinct rather than by following directions. I clung to the ideal of exploring my own personal unknown. I felt that the less I was influenced by other folders, the more I could forge my own path and distinguish my work. The fact that my designs were unrefined didn't bother me. It was not the final outcome of the models that was so important to me, but rather the process of designing them that I loved most.

I mostly folded imaginary creatures. My folding premise was that anything could be made into a creature, and my designing method was to keep folding a piece of paper any which way until there were numerous appendages sticking out. No matter where on the paper the appendages were or how many there were, I would turn them into arms, legs, or wings—and then, presto, I would have a creature. Sometimes it would even turn out to be an almost recognizable creature!

One day when I was in tenth grade I was in a bookstore and discovered the book *Folding the Universe from Angelfish to Zen* by Peter Engel. After buying Engel's book my whole self-protective philosophy changed. I decided that I could better forge my own path if I let myself use the origami tools used by the experts. I bought all the advanced books I could find and studied the different authors' methods of folding. On the whole, these books expanded my repertoire of folding. It was no longer enough to just design a new way to fold a certain subject; the new challenge was to come up with new subjects that had never been folded. I forced myself to branch out as far as I could beyond animal themes and instead try to fold scenes, ideas, and symbols.

One of my main strategies was to try to fold already commonly folded subjects such as cranes, hearts, and people, using only part of the paper so that I could use the rest of the paper to fold some sort of scene surrounding them. For instance, after discovering how to fold a person using only two corners of the square, I was able to apply this method to make a whole variety of models of people doing things. This designing method enabled me to reach outside the existing bounds of origami and define my own style. It also enabled me to put a little bit of my personality into my origami models. Some common themes that would show up in my models were the ridiculously extreme (a 25-headed crane), a ridiculous oxymoron (*Surfer on a Still Lake*), and of course, the just plain ridiculous (*Person Stranded on a Desert Isle, Watching the Sunset*).

Once I've chosen an idea, I find that certain techniques help me actually design a model. One technique I use, before making any folds, is mapping out on a square the points from which each appendage will come when the model is folded. The goal is to plot the points of the appendages on the paper so that the model will waste the least paper and come out as large as possible. For instance, when I design an animal model, I usually plot the head and the tail at opposite corners of the square because for most animals the head is at a point furthest from the tail.

Pinnacle Logo *by Jeremy Shafer, folded from one uncut square sheet of paper. Photo courtesy of Jeremy Shafer.*

Once I have a general idea about where on the square each appendage of my model will come from, the next important technique is to isolate those points. Isolating points involves attaching the points with creases and then using those creases to fold the model in such a way that the points stick out. The easiest way to fold the model so that the points stick out is to start with a standard origami base and then try to adapt it to fit the model.

Designing origami for me has always been a form of self-expression. The process of teaching and diagramming my own models, and even following other people's diagrams, has been a way for me to connect with other folders. Even more satisfying than designing origami models has been sharing them with other people. That why I feel extremely fortunate that my folding path eventually lead me out of the closet into a far greater, much brighter world of worldly origami.

ACTIVITY

8

Folding a Regular Pentagon

o far, you have folded three regular polygons: the square, the regular exagon, and the equilateral triangle. Finding the most elegant way to old the regular pentagon has been a challenge to origami folders. he method you will be learning in this activity was shown by unihiko Kasahara to be one of the more accurate ways to fold the entagon.

ATERIALS NEEDED FOR EACH STUDENT

ne to three squares of origami or patty paper
rigami journal

ROUPING

Vork with a partner. Each person should fold his or her own pentagon.

OLDING INSTRUCTIONS AND QUESTIONS

Vhen you are folding, think about the answers to the folding questions sked at each step. When you have finished folding, answer these uestions in your origami journal.

1 Fold along the diagonal.

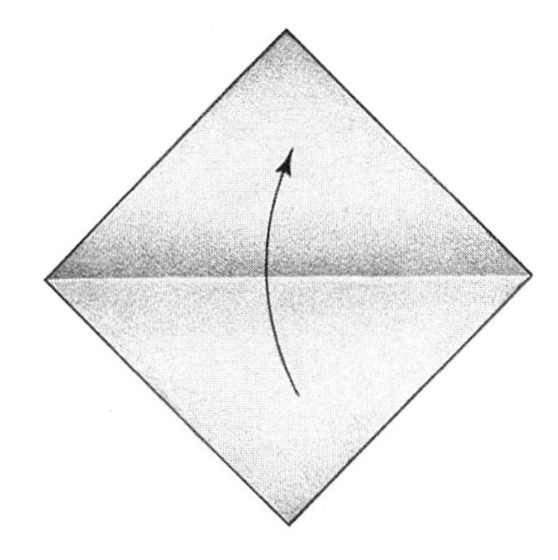

2 Fold in half and unfold.

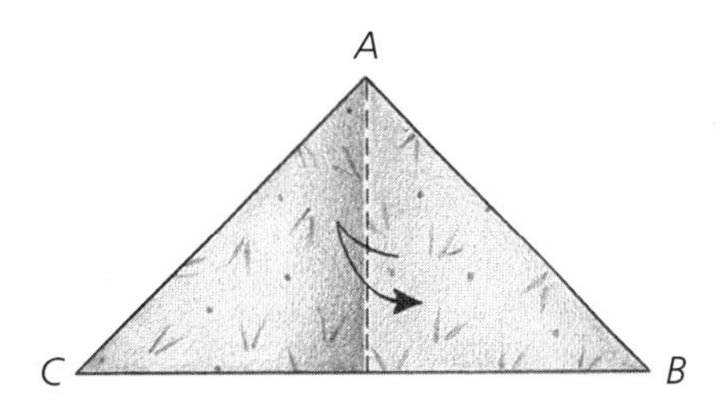

How does the area of each of the right triangles in the diagram above compare to the area of the original square?

3 Mark the midpoint of segment *AB* with a crease by folding and unfolding. Call this midpoint *D*.

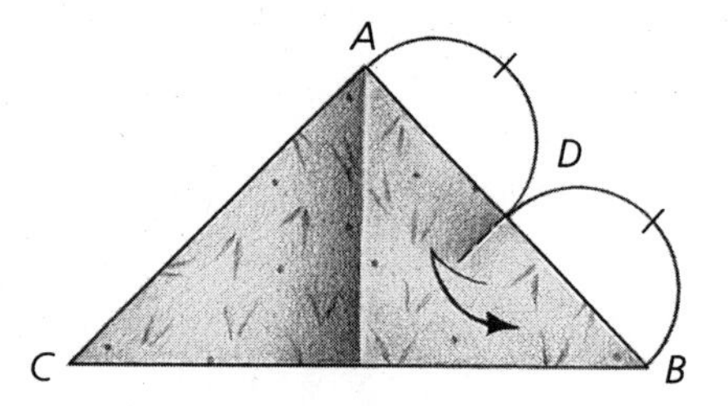

4 Mark the midpoint of segment *AD* with a crease by folding and unfolding. Call this midpoint *E*.

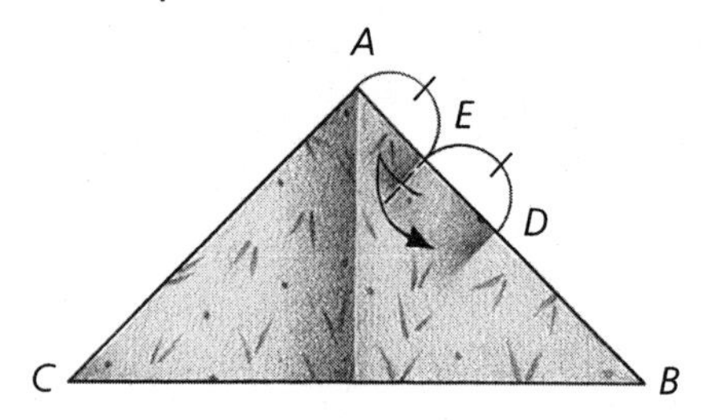

5 Fold and unfold through points *E* and *F*.

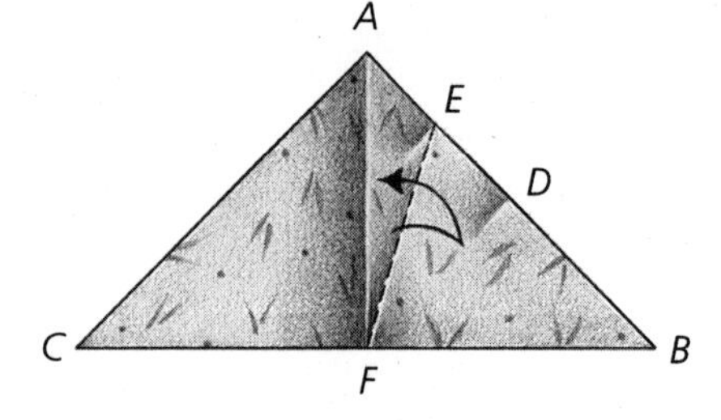

6 Fold to bisect ∠*EFB*.

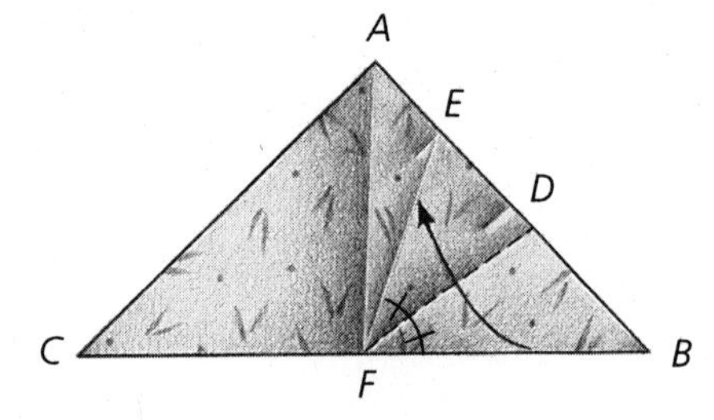

7 Fold to bisect ∠*CFG*.

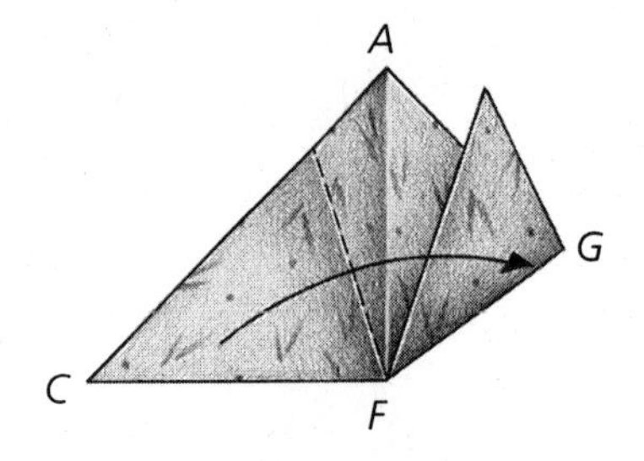

8 Fold the top flap back along segment *FE*.

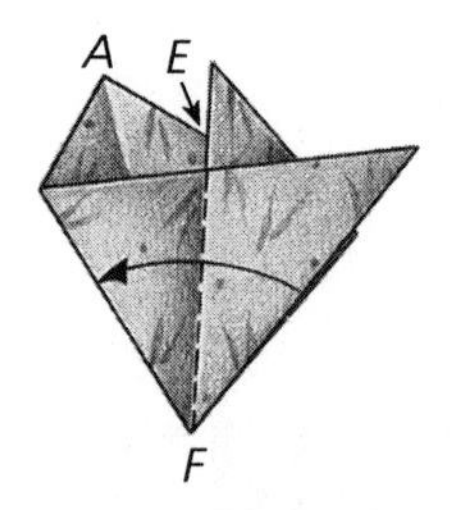

9 Fold all layers on the right side to the back through segment *FE*.

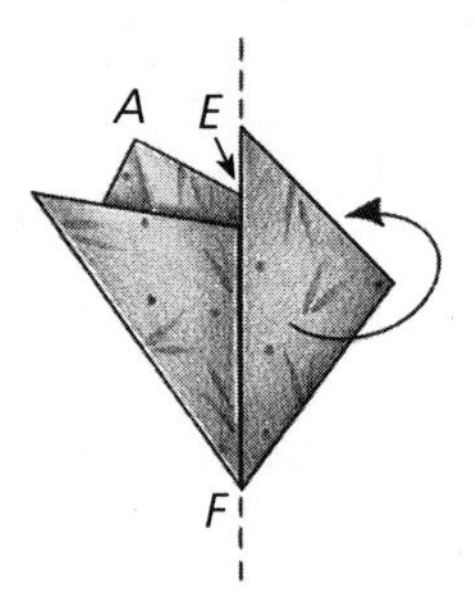

10 Fold all layers. The fold should go through point *G* and be perpendicular to the left side of the figure.

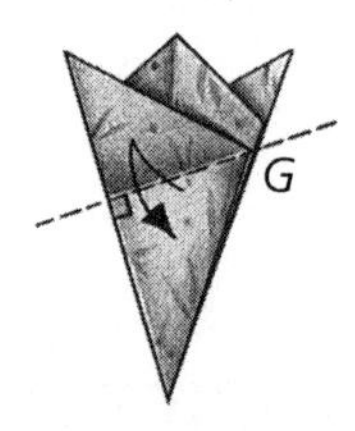

What kind of triangle is formed on each layer?

11 Unfold and open the paper. You have folded a regular pentagon.

EXPLORING YOUR MODEL

Record your answers to the exploring questions in your origami journal.

1. Use your folded pentagon to explore its symmetry. Be sure to add this regular polygon to your chart.
   a. Determine how many lines of symmetry the pentagon has. Describe these lines of symmetry.
   b. Describe the rotational symmetry of the regular pentagon.
2. Make a conjecture about the symmetries of a polygon with n sides. Try your conjecture on the equilateral triangle, square, and regular hexagon.
3. Draw a sketch of your unfolded square. Find the measure of each angle determined by the folds, and write the measures on your sketch.
4. Do the folds in the pentagon lie on the diagonals? Explain why or why not. How would you describe the folds in the pentagon?
5. Use a straightedge to draw the diagonals of the pentagon.
   a. How many diagonals are there? Record this result on your chart.
   b. Describe the figure formed by all of the diagonals.

A PENTAGON STAR

The traditional star that Betsy Ross used on the legendary first flag of the United States is a five-pointed star. Refold one of your pentagons. In the last step, instead of folding perpendicular to the side, make a slanted fold and then cut along the fold. Unfold and you will have a star like the one on the United States flag. You may have to experiment a bit before you get the result you want.

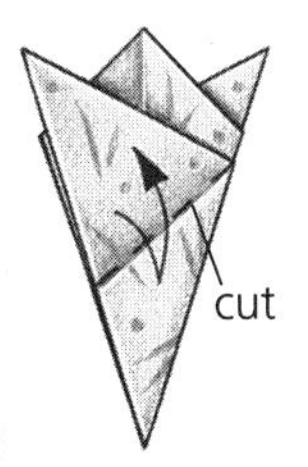

ANOTHER METHOD FOR FOLDING A PENTAGON

The method shown below for folding a pentagon is called the *approximation method.* Try it to see which method you prefer.

1 Fold your square in half.

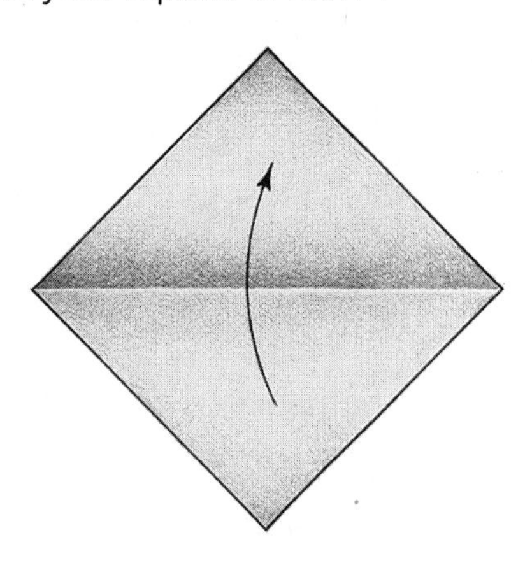

2 Fold in half and pinch to mark the midpoint *D*. Unfold.

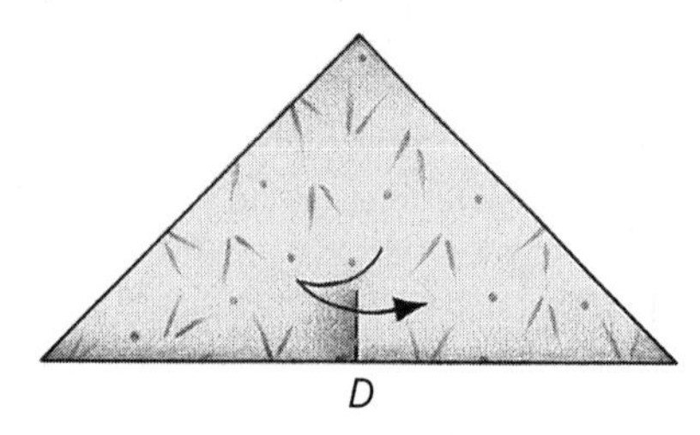

3 Make a fold that pivots on *D*. The goal is to make the fold so that the measure of the angle indicated by $2X$ (as shown in the diagram for step 4) is twice that of the angle indicated by X.

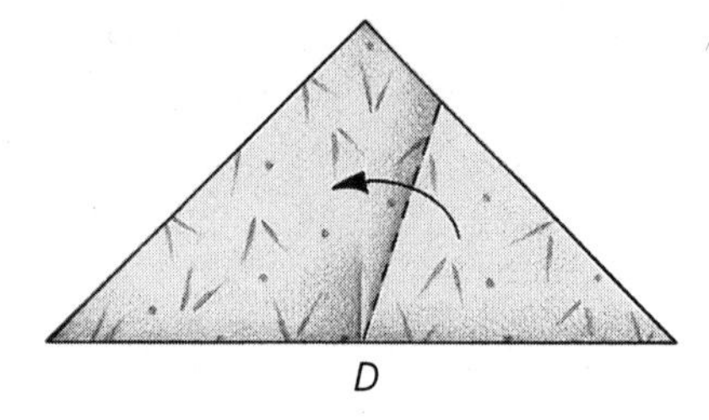

4 Keep adjusting the fold until you think you have the desired relationship for angle measures X and $2X$.

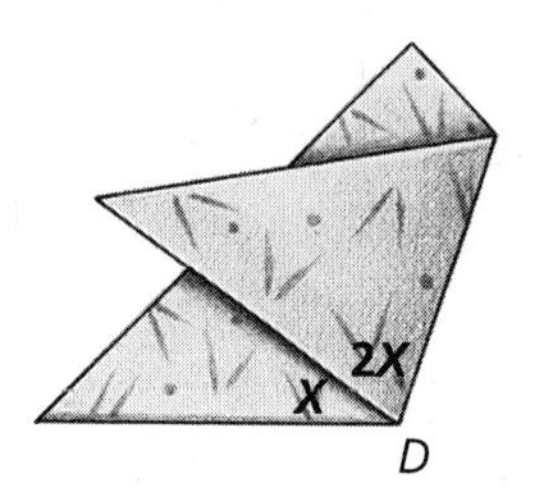

5 Make a fold that bisects the angle with measure $2X$.

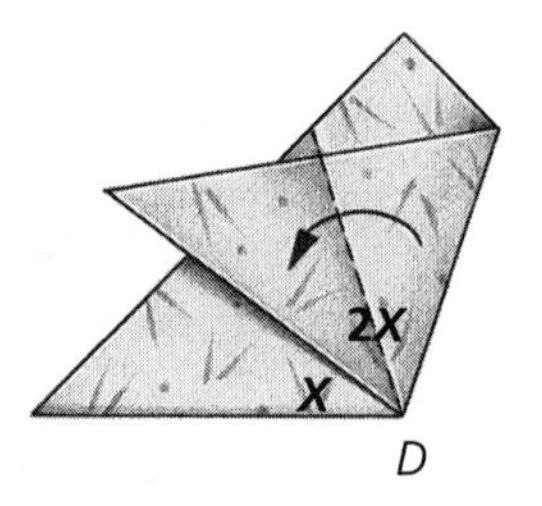

6 Fold back along the dashed line.

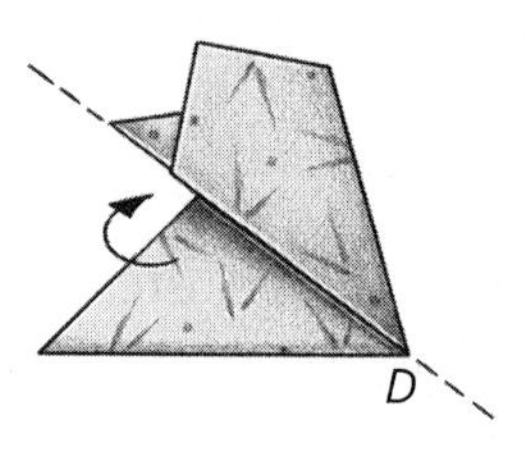

7 If the edges don't match, make adjustments.

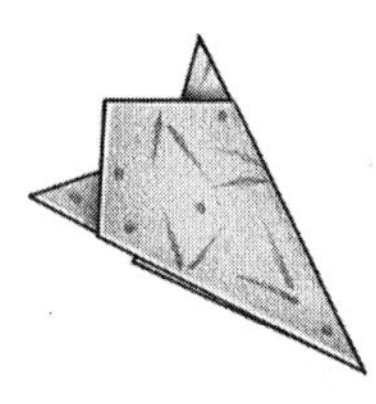

8 Turn the model over and make a fold perpendicular to the left side of the figure.

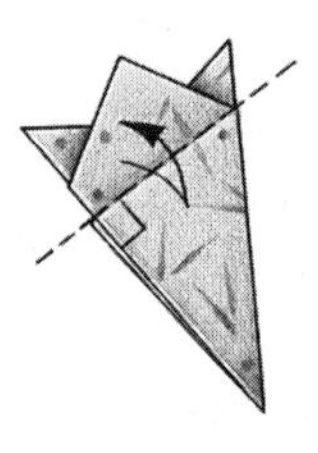

Explain why this approximation method works.

Folding a Regular Pentagon

OBJECTIVES

To fold a regular pentagon
To explore the symmetries of the regular pentagon
To find patterns in the symmetries of *n*-gons

MATERIALS NEEDED FOR EACH STUDENT

One to three squares of origami or patty paper
Student worksheet pages—one copy for each pair of students
Origami journal

TEACHER MATERIALS

Large square for demonstration

TIME

One class period

GROUPING

Students work with a partner. Each student should fold his or her own regular pentagon.

GENERAL INSTRUCTIONS

In this activity, students learn how to fold a pentagon using two different methods. (You can omit the second method if you don't have time for it.) Students explore the reflection and rotational symmetry of the pentagon, and they use their knowledge of geometry to find the measures of the angles formed by the folds. The folding sequences in this activity are somewhat more difficult than those in previous lessons. Be prepared to demonstrate the folds. Encourage students to work together and help each other.

Another interesting folding exercise related to pentagons is the regular pentagonal knot on pages 74–75 of Kasahara's *Origami Omnibus.*

ANSWERS

Folding Questions

1. [No question.]
2. The area of the large triangle is one-half the area of the original square. The area of each smaller triangle is one-fourth the area of the original square.

3–9. [No questions.]

10. Right scalene triangles.
11. [No question.]

Exploring Your Model

1. a. The lines of reflection symmetry are indicated by the folds on the students' pentagons. The regular pentagon has five sides and five lines of reflection symmetry.

   b. The regular pentagon has 5-fold rotational symmetry (72° rotational symmetry). To discover its rotational symmetry, students can trace the pentagon, place the tracing over the original, and rotate the tracing around the tip of a pencil located at the center of the pentagon.

2. Have students make further generalizations about the symmetries of *n*-gons. Review their conclusions that the regular "five-gon" (pentagon) has 5-fold rotational symmetry. They should remember that a regular "six-gon" (hexagon) has 2-, 3-, and 6-fold rotational symmetry. A more sophisticated generalization about the rotational symmetry of regular *n*-gons is that if *n* is prime, then the *n*-gon has *n*-fold rotational symmetry, and if *n* is composite, then the *n*-gon has *n*-fold rotational symmetry and "all the factors of *n*"-fold rotational symmetry.

3.

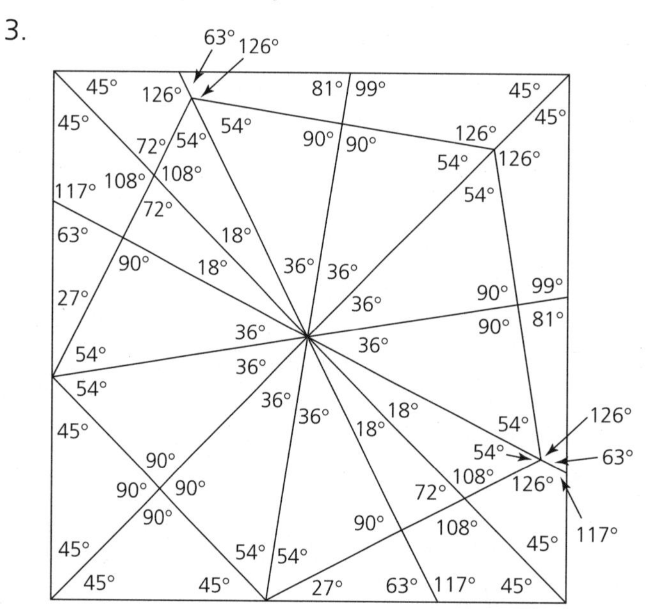

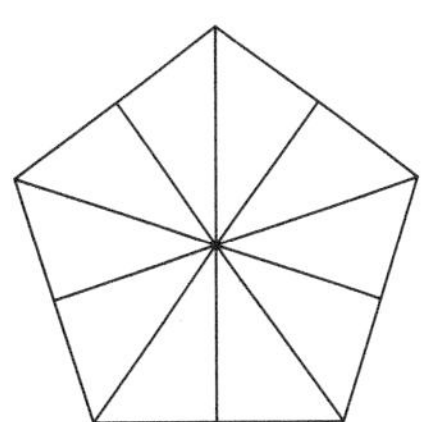

4. None of the folds in the pentagon are diagonals, because diagonals must connect nonconsecutive vertices. The folds do show the central angles as well as the angle bisector of each central angle.
5. a. There are five diagonals.
   b. The diagonals form a five-pointed star.

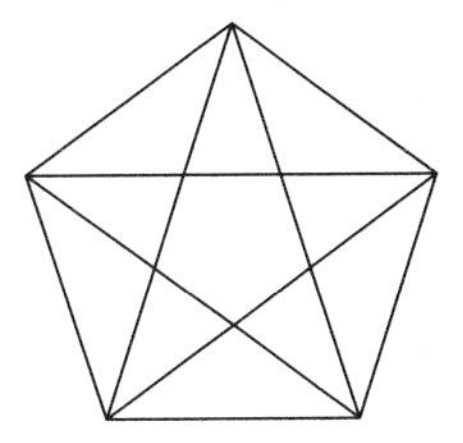

A Pentagon Star

The correct angle is 126°.

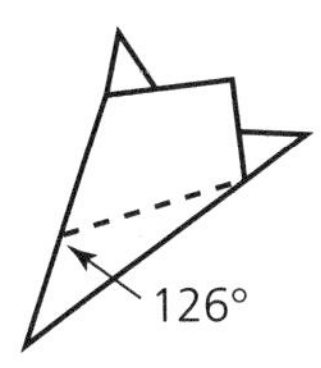

Another Method for Folding a Pentagon

1–7. [No questions.]

8. The approximation method of folding a regular pentagon works because the folding involves dividing the area of half a square into five equal regions. When the square is opened, you can observe the ten divisions around the central angle and the five equal edges of the pentagon.

ACTIVITY 9

Octagon Star

In this activity you will be introduced to unit, or modular, origami. Japanese folder Tomoko Fusè has created many unit origami models. Her work has inspired many other folders to use origami to model polyhedra. To create a unit origami model you build two or more units that are exactly alike. Then, without using glue, you put the units together to form the final model. Robert Neale first created the model in this activity, which he calls the *Pinwheel-Ring-Pinwheel.* The unit in this model is very simple to fold. You will probably find it quite easy to assemble the model as well. But the end result will be very striking, and it even transforms itself from an octagon ring to an octagon star.

MATERIALS NEEDED FOR EACH STUDENT OR PAIR OF STUDENTS

Eight squares of origami paper—four each of two different colors or two each of four different colors
Origami journal

GROUPING

Work individually or with a partner.

FOLDING INSTRUCTIONS AND QUESTIONS

When you are folding, think about the answers to the folding questions asked at each step. When you have finished folding, answer these questions in your origami journal.

1 Fold the paper in half and unfold.

2 Rotate the paper 90°. Fold in half and unfold.

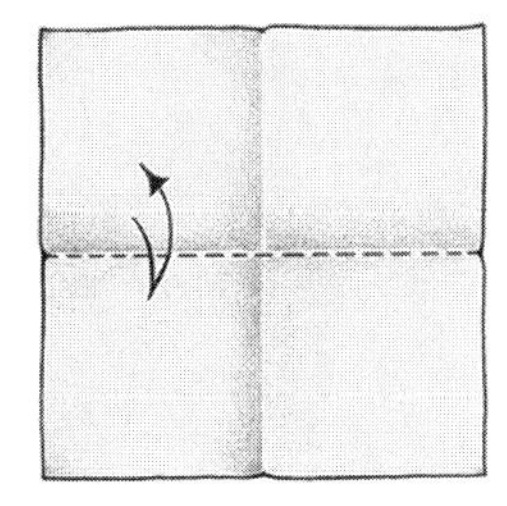

3 Fold the top two corners down.

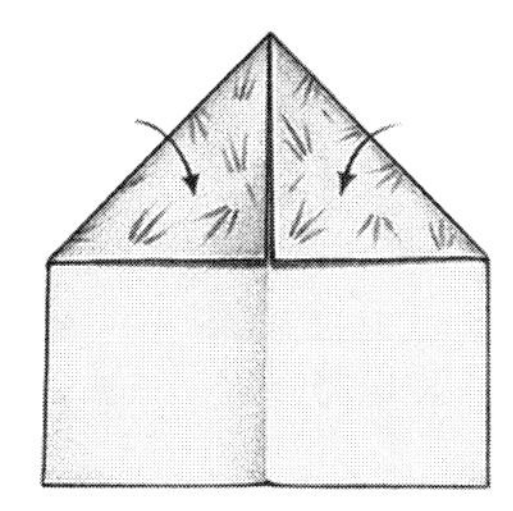

How does the area of the pentagon above compare to the area of the original square?

4 Fold the white sides together on the diagonals indicated in the diagram. Crease only to the middle of the square.

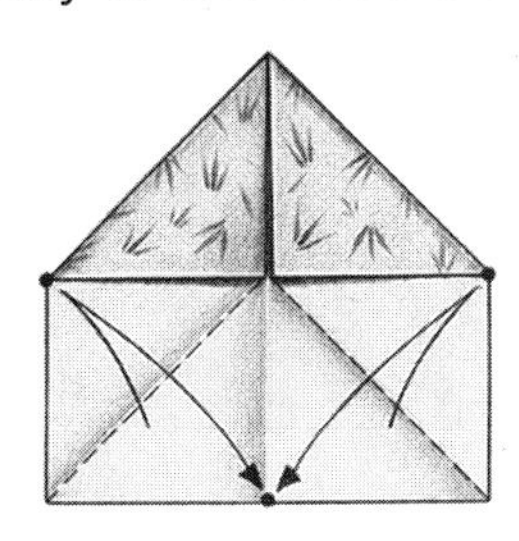

5 Rotate the piece 90° and fold in half.

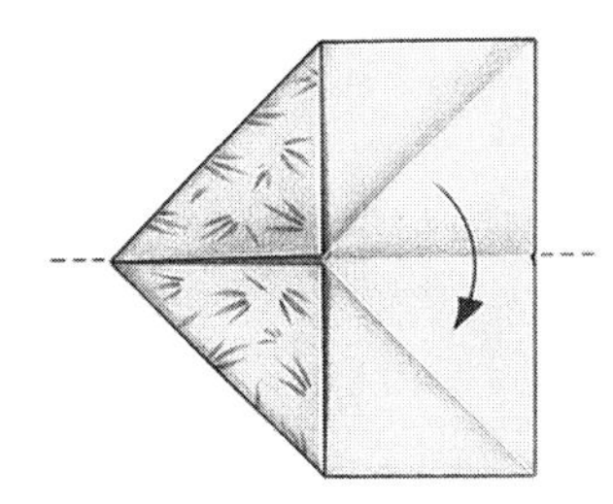

6 Push the fold to the inside so that a parallelogram is formed.

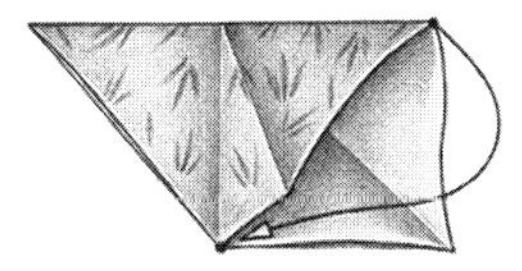

7 You have completed one of the units for the octagon star. Now make seven more units just like this one.

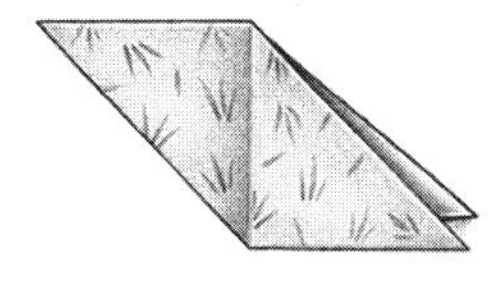

What are the measures of each angle of the parallelogram? Explain how you know.

Remember to answer each of the folding questions in your origami journal.

ASSEMBLY INSTRUCTIONS

1 Pick up two different-colored units. Insert the short side of one of the units between the long folded edges of the other unit.

2 Hook the parallelograms together by tucking in the "tails."

3 Continue connecting the parallelograms together until you have formed an octagon.

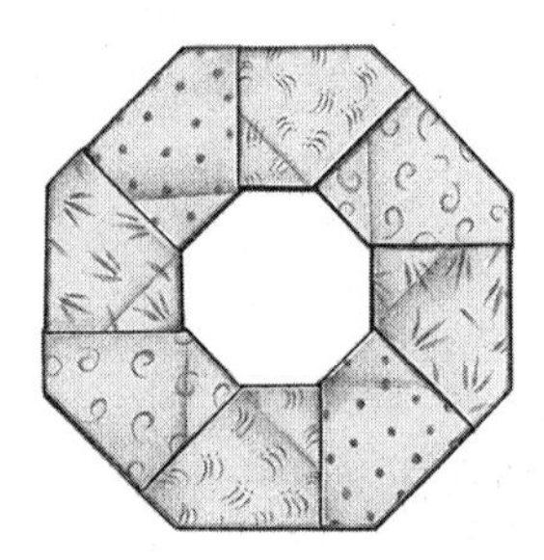

4 Slide the pieces of the octagon togethe until an eight-pointed star is formed. If your units don't slide easily, then check to see that when you tucked in the last two "tails" both were folded over the edges and not over the center piece.

EXPLORING YOUR MODEL

Answer the exploring questions in your origami journal.

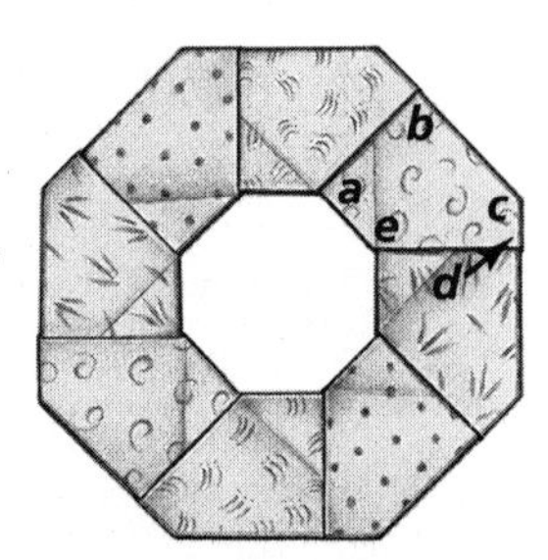

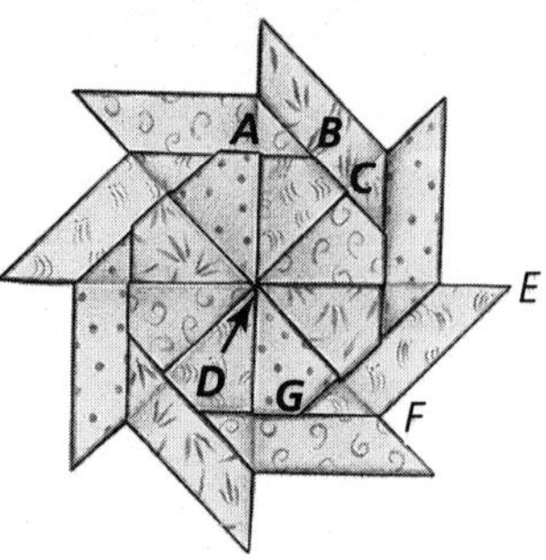

1. Without measuring, determine the measure of angles *a*, *b*, *c*, *d*, and *e* in your octagon ring.
2. Determine the measure of each of the following angles in the octagon star: $\angle DAB$, $\angle ABC$, $\angle BCD$, $\angle GFE$, and $\angle E$. What kind of quadrilateral is *ABCD*?
3. Describe the rotational and reflection symmetries of both the octagon ring and the octagon star. Are the symmetries the same or different? Explain.
4. Describe the relationship between the parallelogram in the completed octagon star and the original unit you made to build the star.

Octagon Star

OBJECTIVES

To introduce students to unit origami
To explore the symmetry of an octagon ring and an octagon star
The find the measures of angles in a parallelogram
To observe that two parallelograms with the same angle measure are not necessarily similar

MATERIALS NEEDED FOR EACH STUDENT OR PAIR OF STUDENTS

Eight squares of origami paper—four each of two different colors or two each of four different colors
Two large units to demonstrate how pieces fit together
Student worksheet pages—one copy for each pair of students
Origami journal

TEACHER MATERIALS

Several large squares for demonstration

TIME

One class period

GROUPING

Students should work individually or with a partner.

GENERAL INSTRUCTIONS

Students really enjoy making this model, so it is a good activity to use to introduce unit origami. Each student will probably want to make his or her own model. Some students might enjoy making a half-size model. The model is also very effective if done in eight different colors.

ANSWERS

Folding Questions

1–2. [No questions.]

3. The area of the pentagon is $\frac{3}{4}$ that of the original square.

4–6. [No questions.]

7. The acute angles each have a measure of 45°, and the obtuse angles each measure 135°.

Exploring Your Model

1. $m\angle a = 90°$, $m\angle b = 90°$, $m\angle c = 135°$, $m\angle d = 90°$, and $m\angle e = 135°$.
2. $m\angle DAB = 90°$, $m\angle ABC = 135°$, $m\angle BCD = 90°$, $m\angle GFE = 135°$, and $m\angle E = 45°$. Quadilateral *ABCD* is a kite.
3. Answers will vary. Because of the orientation of the folds, neither figure has reflection symmetry. If color is not taken into account, or if only one color is used, the octagon ring has 8-fold rotational symmetry. The octagon star has 4-fold rotational symmetry if you consider the colors in a two-color model. It has 8-fold rotational symmetry if you don't consider the colors.
4. The measures of the corresponding angles are equal, but the parallelograms are not similar because the ratios of their side lengths are not proportional.

ACTIVITY

10

Star-Building Unit

In this activity you will learn how to fold a traditional origami unit that can be used to create many different polyhedra. The folding is easy and will become so mechanical that you will be able to do it without thinking. David Masunaga has explored many different polyhedra that can be made from the star-building unit. You will be exploring some of these polyhedra in the activities that follow.

MATERIALS NEEDED FOR EACH STUDENT

One or two squares of origami or patty paper
Origami journal

GROUPING

Work with a partner in your group. Each student should assemble his or her own star-building unit.

REMINDER

Save your completed units for the next activity.

FOLDING INSTRUCTIONS AND QUESTIONS

When you are folding, think about the answers to the folding questions asked at each step. When you have finished folding, answer these questions in your origami journal.

1 Start with the white side of the paper facing up. Fold the paper into two congruent rectangles and unfold.

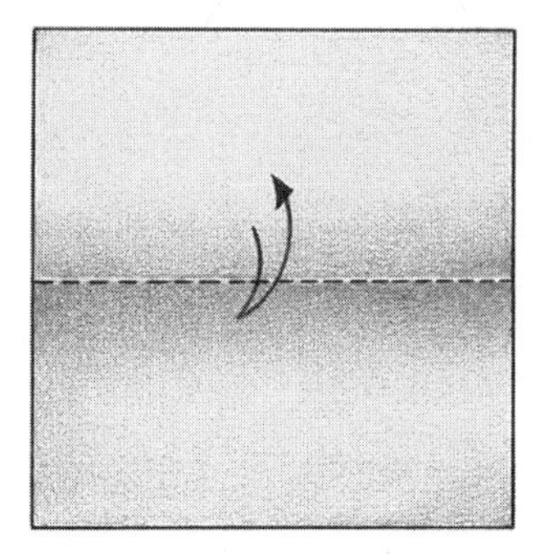

What can you say about the area of each small rectangle compared to the area of the square?

2 Fold each small rectangle in half lengthwise and unfold.

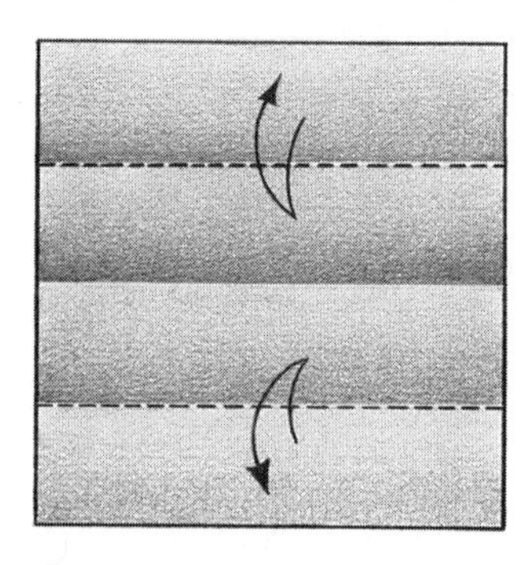

a. *What is the area of each of the smallest rectangles compared to the area of the square?*

b. *Describe the relationship among the three folded lines on your paper.*

3 Fold the lower corner up.

a. *What kind of triangle have you folded?*

b. *What are the measures of its angles?*

4 Rotate the paper 180° and repeat step 3.

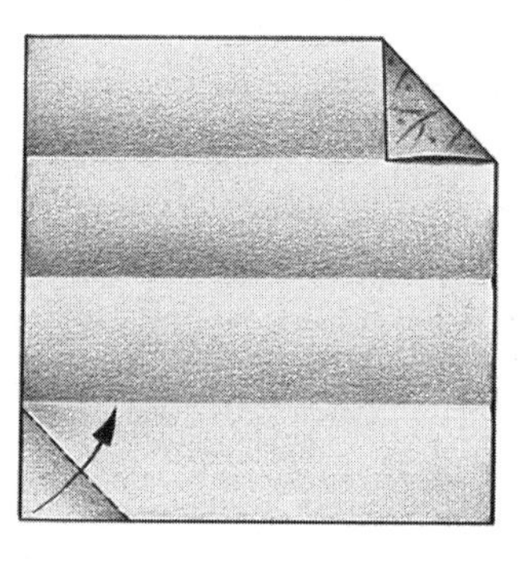

a. *What is the name of the six-sided polygon you have made? Is it a regular polygon? Why or why not?*

b. *Put your finger on the center point of the paper and rotate the figure 180°. Explain why you can say that this figure has 180° rotational symmetry.*

Fold to bisect the 45° angle as shown. This fold is known as the "paper airplane" fold. Be sure to keep the vertex point as sharp as possible.

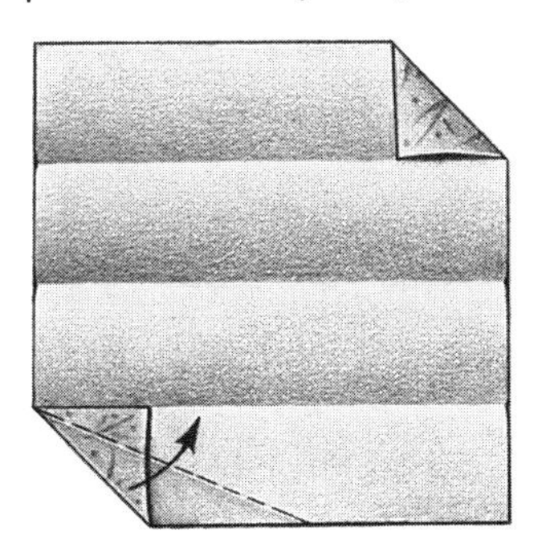

Rotate the paper 180° and repeat step 5.

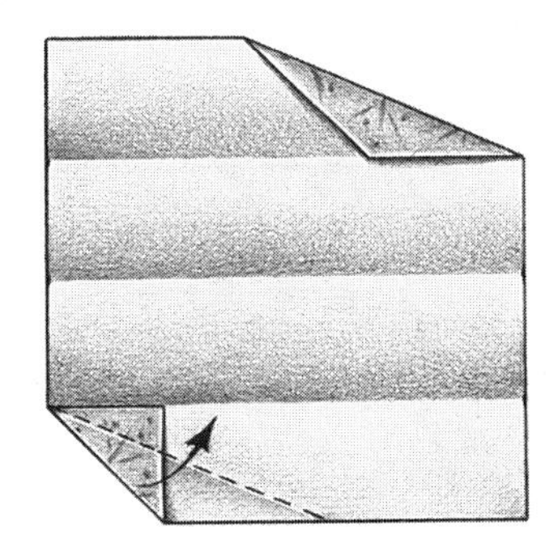

a. What kind of triangle is this final folded triangle?

b. What are the measures of its angles?

7 Refold along the existing parallel line segments *AB* and *CD*.

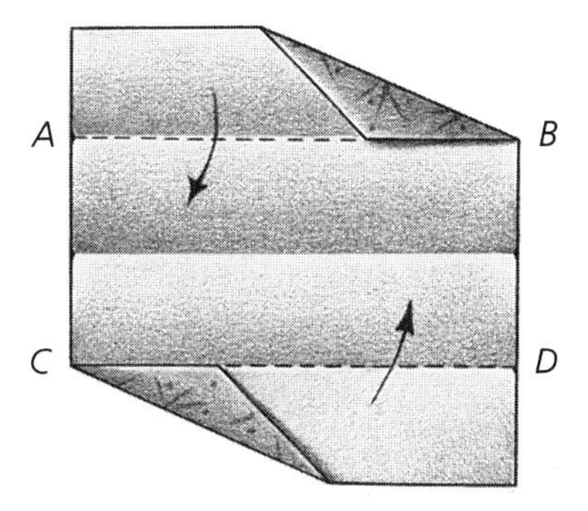

8 Starting from the lower right-hand corner, fold a large isosceles triangle so that point *F* lies on point *G* and point *H* is a vertex.

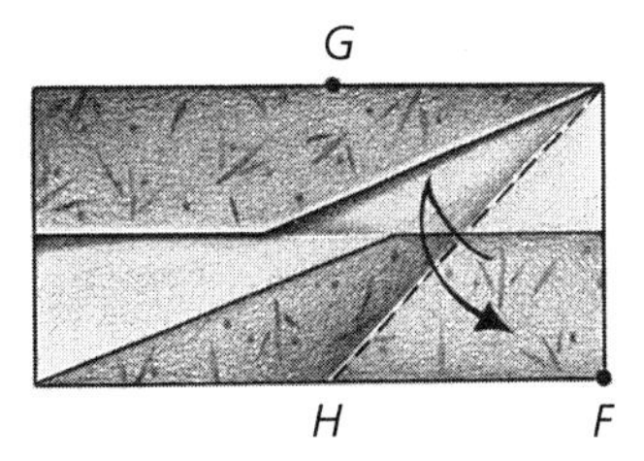

a. Look for the two congruent triangles in the diagram above. What kind of triangles are these?

b. What kind of polygons are the two congruent patterned figures?

9 Rotate the paper 180° and repeat step 8.

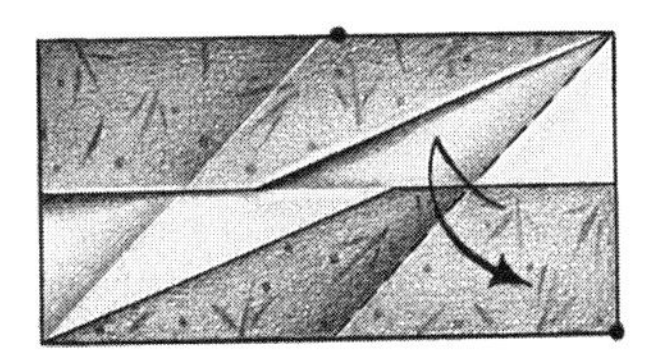

10 Tuck each flap into a pocket, making sure the corners lie flat when inserted.

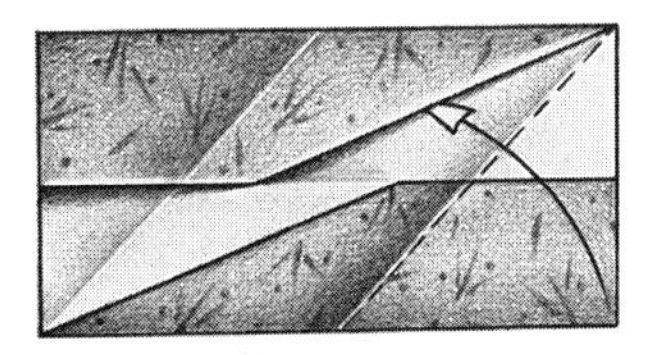

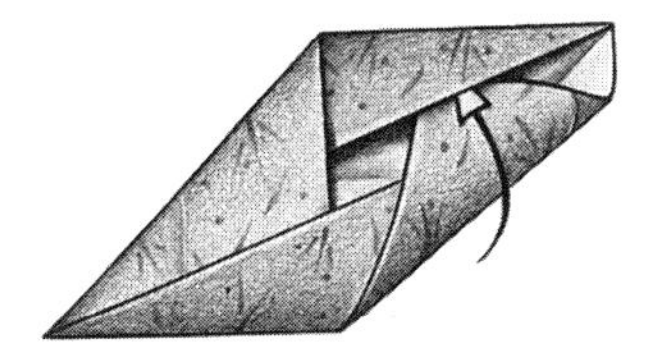

11 The star-building unit is now complete. Your figure should now look like the one below. The size of the white space on your unit, as shown in the picture below, reveals how accurate your folds have been. A small white space won't make much difference, but a gaping hole might.

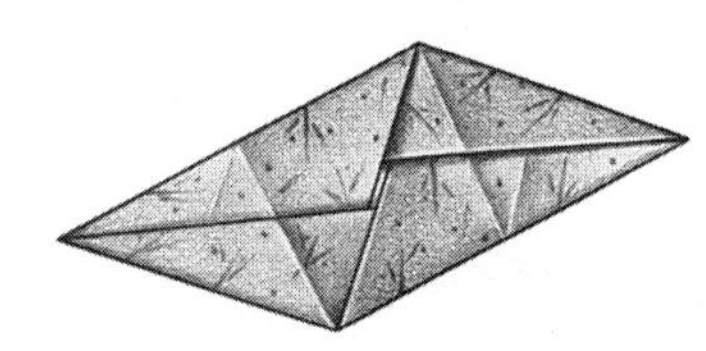

12 Here's how to fold the unit for easy storage. Turn the figure over.

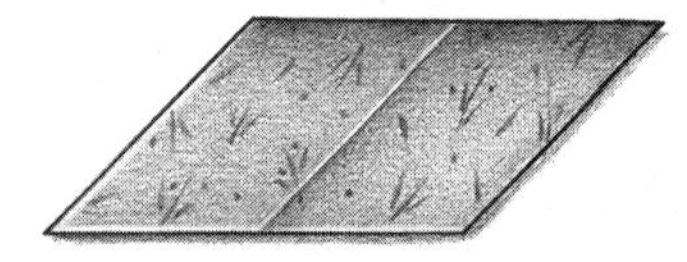

a. *Name the quadrilateral formed.*

b. *What are the measures of the angles of this quadrilateral?*

13 Fold the top right acute angle so that its vertex lies on top of the top left obtuse angle.

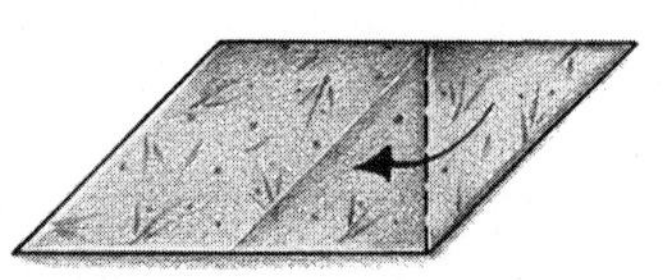

a. *What kind of triangle is formed by your fold?*

b. *What can you say about the area of the triangle compared to the area of the quadrilateral?*

14 Rotate the figure 180° and repeat step 13.

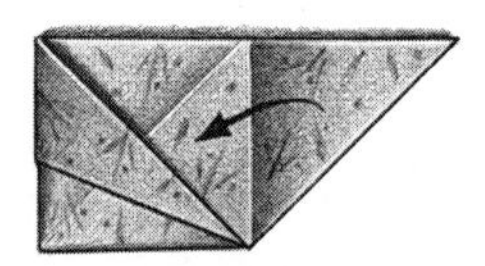

15 Leave the unit in its square shape, and put it in a safe place.

How does the area of the final square compare to the area of the quadrilateral in step 12?

Remember to answer each of the folding questions in your origami journal and save your completed star-building units for the next activity.

EXPLORING YOUR MODEL

Record your answers to the exploring questions in your origami journal.

1. Unfold your star-building unit to step 11. What degree of rotational symmetry does your star-building unit have? Write about how you know.

2. Look at your partially unfolded unit. Name as many different kinds of triangles as you can find. Record your findings in your journal.

3. Most traditional origami is mirror symmetric. If you unfold a figure, the fold lines have mirror symmetry. What sets unit origami apart from other forms of origami is that most of the units have rotational symmetry rather than mirror symmetry. For this reason, you will see the ↶ symbol quite often in constructions. The accompanying sketch shows an unfolded star-building unit. Explain why this diagram has rotational symmetry and is not mirror symmetric.

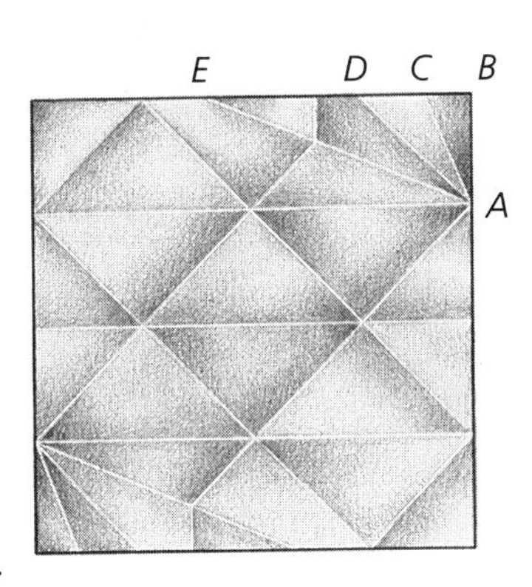

4. Refer to the diagram in question 3.

   a. Find the measures of each of these angles: $\angle CAB$, $\angle DAB$, and $\angle EAB$.

   b. What kind of triangles are ΔABC, ΔABD, ΔADE, and ΔABE? For each of these, be as specific as you can.

   c. Which two triangles named in question 4b are similar? Explain how you know.

5. You could say that this is a left-handed star-building unit because the first triangle you folded was in the bottom left-hand corner (see step 3). What would happen if the first triangle you folded was in the bottom right corner? Try it and see. Compare the resulting figure with the original unit you made. How are the two units alike? How are they different?

6. In unit origami you put together several units to make a model. Experiment in your group and see if you can discover how to put two or more units together. (Be sure to use only right-handed or left-handed units when making your model.)

Star-Building Unit

OBJECTIVES

To fold the star-building unit
To understand the function of a basic unit
To compare areas of polygons
To identify different kinds of triangles and the measures of their angles
To explore rotational symmetry

MATERIALS NEEDED FOR EACH STUDENT

Two squares of colored origami or patty paper
Student worksheet pages—one copy for each pair of students

TEACHER MATERIALS

Overhead projector and waxed paper, or large paper square for demonstration
Completed star-building unit

TIME

One class period

GROUPING

Work with a group of four. Each student should assemble his or her own star-building unit.

REMINDER

Students need to save any star-building units they fold in this activity for the next activity.

GENERAL INSTRUCTIONS

Encourage students to work together and help each other figure out the folding sequence. This is a fairly easy unit to fold, so students should not have much difficulty. You might want to allow students to fold their first unit and then go back and fold another unit while answering the folding questions. Each student will need four star-building units for the next activity.

In question 5, students explore right- and left-handed star-building units. The notion of right- and left-handedness applies to many contexts. It is addressed by Martin Gardner in his book *The Ambidextrous Universe* and by Lewis Carroll in *Alice Through the Looking-Glass.* In organic chemistry, chemicals such as sucrose have right- and left-handed versions. In hospitals, knobs have right- and left-handed turns to minimize errors.

ANSWERS

Folding Questions

1. The area of each small rectangle is one-half the area of the square.
2. a. The area of each of the smallest rectangles is one-fourth the area of the square.
   b. The three folded lines are parallel and congruent segments.
3. a. The folded triangle is an isosceles right triangle.
   b. The measures of the angles are 45°, 45°, and 90°.
4. a. The six-sided polygon is a hexagon. It is not a regular hexagon because the sides have different lengths.
   b. When you pivot the paper 180° around its center point, the rotated figure looks exactly the same as the original figure.
5. [No question.]
6. a. The final folded triangle is an isosceles, obtuse triangle.
   b. The measures of the angles are 22.5°, 22.5°, and 135°.
7. [No question.]
8. a. The two congruent triangles are right, scalene triangles.
   b. The two congruent patterned figures are right trapezoids.

9–11. [No questions.]

12. a. The quadrilateral is a parallelogram.
    b. The measures of the angles are 45°, 135°, 45°, and 135°.
13. a. The triangle formed is an isosceles right triangle.
    b. The area of the folded triangle is one-fourth the area of the parallelogram.
14. [No question.]
15. The area of the square is one-half the area of the parallelogram.

Exploring Your Model

1. The star-building unit in step 11 has 180° rotational symmetry. This is because when you rotate the figure 180° about its center point, the rotated figure looks exactly the same.
2. Students should be able to find right isosceles, isosceles, right scalene, and scalene triangles visible in the basic star-building unit.
3. The figure has 180° rotational symmetry because if you rotate it around its center point, it will look exactly the same. The figure does not have mirror symmetry because there is no way to fold it so that the small right triangle in the upper right-hand corner lies on top of the small right triangle in the lower left-hand corner.
4. a. $m\angle CAB = 22.5°$, $m\angle DAB = 45°$, $m\angle EAB = 67.5°$.

   b. ΔABC is a scalene, right triangle; ΔABD is an isosceles, right triangle; ΔADE is an isosceles, obtuse triangle; ΔABE is a right, scalene triangle.

   c. ΔABC and ΔABE are similar. The measures of their corresponding angles are equal.
5. A right-handed unit and a left-handed unit are both parallelograms, and they are congruent to each other. If you were to take two left-handed units and place them on top of each other, the same side of each would be facing up. However, if you place a left-handed unit on top of a right-handed unit so that their shapes conform, opposite sides are facing up.
6. Answers will vary. Some groups may figure out how to assemble the triangular hexahedron that is presented in Activity 11. If they do, they should still complete the Exploring Your Model questions when they get to that activity.

Triangular Hexahedron

ı this activity, you will be creating a three-dimensional origami model. The rawings can no longer show every angle of the figure you are making. ou will need to use what you see in the drawing, your common sense, our spatial sense, and your sense of adventure. If trying one thing doesn't vork, try another. Working with your partner or others in your group will elp. You will need to complete Activity 10 before doing this activity.

⁄IATERIALS NEEDED FOR EACH STUDENT

hree different-colored squares of origami or patty paper: light, medium, and dark
)rigami journal
ˌutton
hread or string

ƷROUPING

Vork with a group of four. Each person should assemble ıis or her own triangular hexahedron.

ꝚEMINDER

ƽave your completed model for the next activity.

ASSEMBLY INSTRUCTIONS

Ʒefore assembling the triangular ıexahedron, make three star-building units using papers in three different shades or patterns: light, medium, and dark. Color the squares at the right to track the color numbers as you follow the assembly instructions.

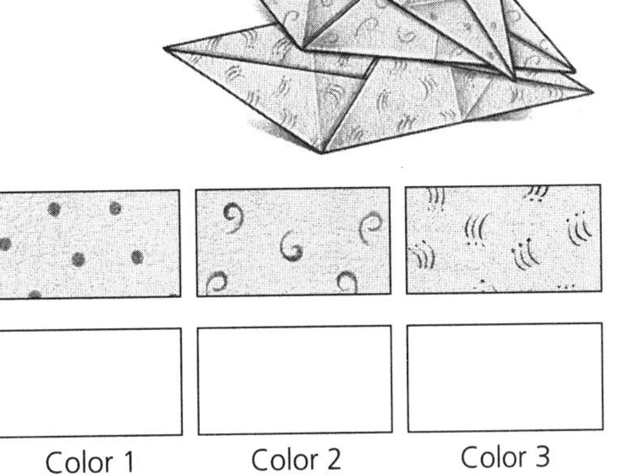

1 To use the star-building units to make three-dimensional shapes, you fit the pieces together by inserting each tab end into a pocket. Each star unit has two tabs and two pockets. Look at one of your units and identify the tabs and pockets.

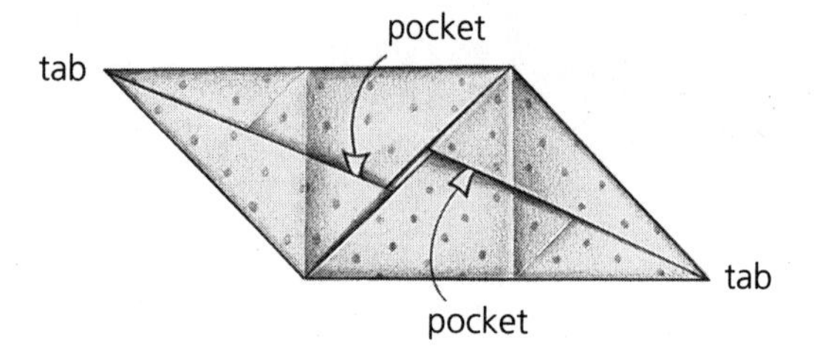

2 Make a fold along the diagonal that contains the white space or small opening on each unit.

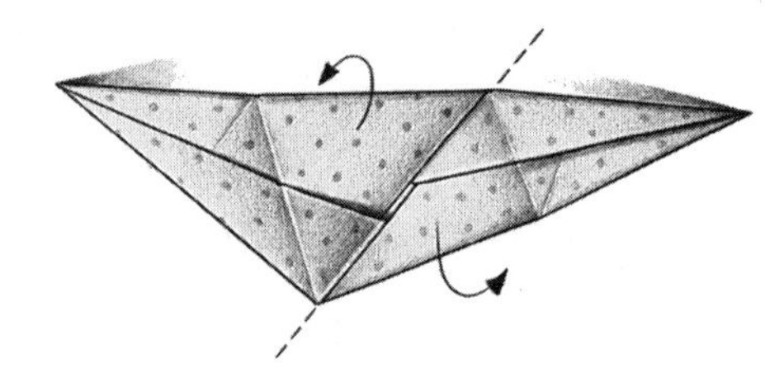

The folded unit should look like this.

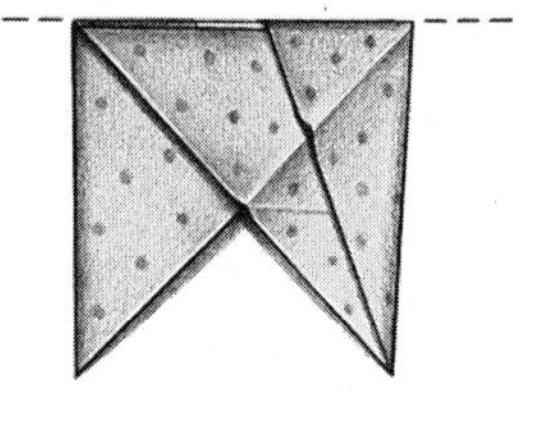

3 Unfold the unit. The fold will be a crease now. Do step 2 for each star-building unit.

4 Fit a tab of the color 2 unit into a pock of the color 1 unit as shown.

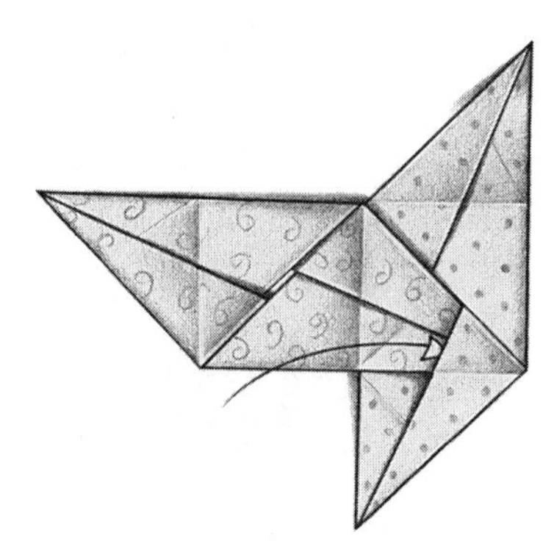

5 Rotate the figure 90° clockwise. Next fi a tab of the color 3 unit into the pocket of the color 2 unit as shown.

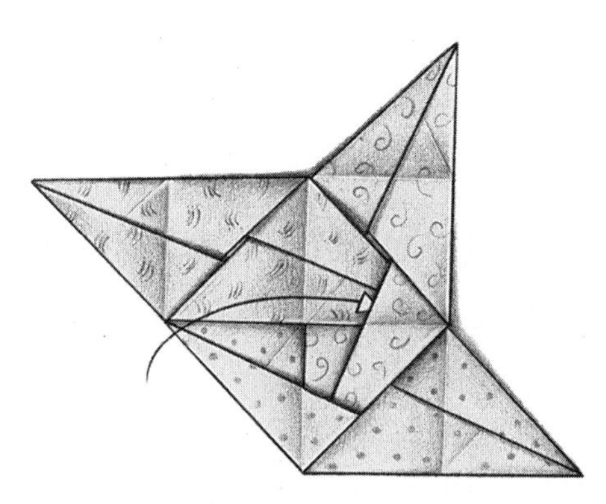

6 So far your piece is still two-dimensiona In this step you will make it three-dimensional. Crease the units along the three mountain folds as shown. Then squeeze the three shapes together, inserting the tab of the color 1 unit into the pocket of the color 3 unit.

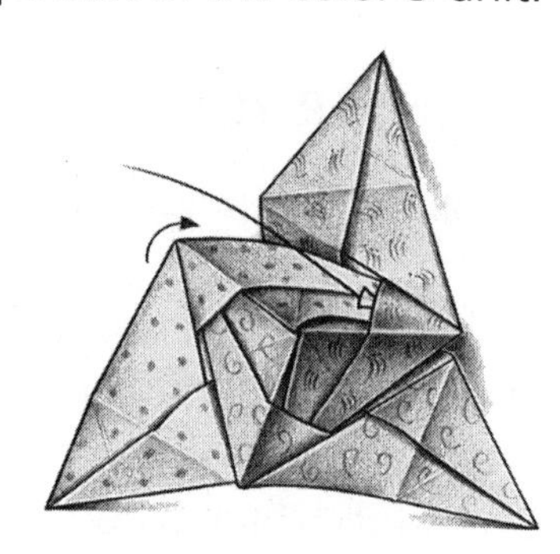

7 Turn the model over.

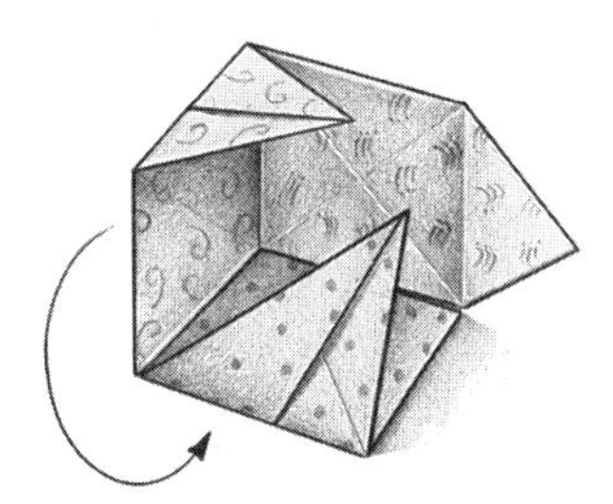

8 Insert the tab of the color 3 unit into the open pocket of the color 1 unit.

9 Insert the tab of the color 1 unit into the pocket of the color 2 unit. Make sure that you leave the tab of the color 2 unit outside the figure.

10 Insert the tab of the color 2 unit into the pocket of the color 3 unit.

11 Now you have a triangular hexahedron.

EXPLORING YOUR MODEL

Record your answers to the exploring questions in your origami journal.

1. When you hold your triangular hexahedron, how many triangles do you see? Can you rotate your hexahedron so that you see a different number of triangles? How many different ways can you look at your triangular hexahedron?
2. Study your model and try to describe the planes of rotational symmetry. How many axes of symmetry are there?
3. Cut out cross section A, provided by your teacher. Place the model halfway through the cross section. Describe what each half of your model would look like if you were to cut the triangular hexahedron with this plane.

4. Cut out cross section B, provided by your teacher. Place the model halfway through the cross section. Describe what each half of your model would look like if you were to cut the triangular hexahedron with this plane.
5. A polyhedron is a solid whose faces are all polygons. What kind of triangles make up the faces of your triangular hexahedron? How many vertices does each face of your triangular hexahedron have?
6. A hexagon is a six-sided polygon. What do a hexagon and a triangular hexahedron have in common? What is different about them?
7. You can describe a polyhedron by telling the numbers of its faces, vertices, and edges. Write a description of your triangular hexahedron. Include the number of faces, vertices, and edges in your description.
8. Do the same number of edges meet at each vertex? Explain.
9. The suffixes of two- and three-dimensional figures have interesting meanings. The suffix *-gon,* as in *hexagon,* means "angle." The suffix *-hedron,* as in *hexahedron,* means "base." A hexahedron is a polyhedron with six faces. Make a table in your origami journal like the one that follows. Predict the names of the other polyhedra on the list.

Number of faces	Name of polyhedron
4 faces	tetrahedron
5 faces	
6 faces	hexahedron
7 faces	
8 faces	
9 faces	
10 faces	

10. The triangular hexahedron is also called a *double tetrahedron* or a *triangular bipyramid.* Explain why the triangular hexahedron might be given these names.

HANGING YOUR MODEL

To hang your model, secure a piece of thread or string to a button. Before closing the model at a vertex, place the button inside the model. In Japanese ornamental decorations, string is threaded through the whole model with a decorative tassel attached to the end. The model's rotational symmetry makes it symmetrical when it is hung.

Triangular Hexahedron

OBJECTIVES

To assemble a triangular hexahedron
To introduce the definition of a polyhedron
To count the faces, vertices, and edges of a polyhedron
To explore the reflection and rotational symmetry of a triangular hexahedron

MATERIALS NEEDED FOR EACH STUDENT

Three different-colored squares of origami or patty paper
Student worksheet pages—one copy for each pair of students
Origami journal
Button
Thread or string

TEACHER MATERIALS

Three large star-building units for demonstration

TIME

One class period

GROUPING

Students should work in groups of four. Each student will assemble his or her own triangular hexahedron.

REMINDER

Have students save their completed models for use in the next activity.

GENERAL INSTRUCTIONS

Some students may find it difficult to make this model by just following the written instructions or even by looking at the diagrams. You may have to demonstrate how to fit the pieces together. Usually at least one student in the group will figure it out, however, and these students should be encouraged to help others in their group. As an extension of this activity you could have the students fold the bird tetrahedron (which is really a triangular hexahedron) on pages 134–135 of *Unit Origami* by Tomoko Fusè. They can then create some of the multi-unit assemblies shown on pages 138–141 of that book.

Depending on your students' abilities, there are several options for having them explore the reflection planes of the model. You can have students create their own cross-section cutouts, you can have students cut out the cross-section diagrams, or you can demonstrate the reflection planes using precut cross sections. If you choose to do a demonstration, you will find that cutting the cross sections out of plastic transparency material works very well. The cross-section diagrams provided work with models made from 6-inch square paper.

If you ask students to create their own cross-section cutout, they may need a polyhedron protractor, or you can have students estimate and use guess-and-check to cut the cross section. Students' cross sections don't need to be exact.

ANSWERS

Exploring Your Model

1. You will see two, three, or four triangles depending on how you hold the model. There are two ways to hold the model to see three triangles, three ways to hold the model to see two triangles, and three ways to hold the model to see four triangles.
2. There are three axes of 2-fold rotational symmetry (180° rotational symmetry). There is one axis of 3-fold rotational symmetry (120° rotational symmetry).
3. Answers will vary. Each half will be a pyramid with a kite base. There are two pairs of congruent triangular faces.
4. Answers will vary. Each half will be a triangular pyramid with an equilateral triangular base. The faces are all congruent triangles.
5. Each face of the triangular hexahedron is a 45-45-90 triangle. Each face has three vertices.
6. A hexagon has six edges and a triangular hexahedron has six faces. A hexagon has two dimensions, and a triangular hexahedron has three dimensions. It is a solid.
7. The triangular hexahedron has six faces, five vertices, and nine edges.
8. At two of the vertices, three edges meet. We say that these vertices have a *valency* of 3 or are 3-valent. At three of the vertices, four edges meet. We say that these vertices have a valency of 4 or are 4-valent.

9. The name of a polyhedron is determined by the number of its faces.

Number of faces	Name of polyhedron
4 faces	tetrahedron
5 faces	pentahedron
6 faces	hexahedron
7 faces	septahedron
8 faces	octahedron
9 faces	nonahedron
10 faces	decahedron

10. A triangular hexahedron looks like two tetrahedra joined with a common base. It can also be seen as two triangular pyramids.

Templates for Cross Sections (6-inch paper)

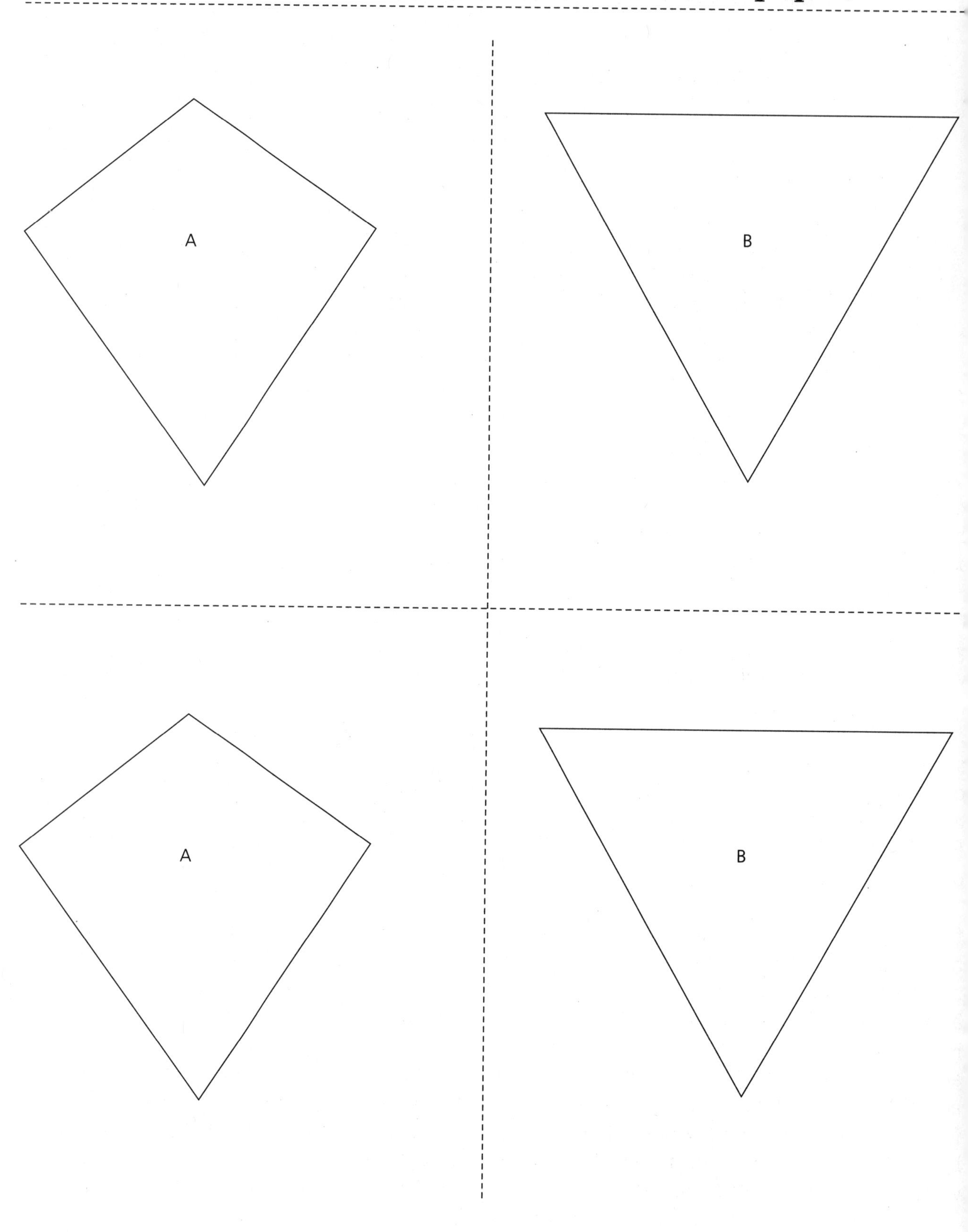

My Experiences with Origami

Robert J. Lang

Robert Lang

Robert J. Lang lives in Pleasanton, California, and is an origami master, the author of six books, and a physicist. He recently collaborated with fellow origami artist Peter Engel on a CD-ROM titled Origami: The Secret Life of Paper, *which has received major computer magazine awards. The following describes Robert's experiences with origami and how origami, mathematics, and science have intertwined in his life.*

I began folding at the age of six. Like many Western folders, my first exposure to origami was through a book, which in my case contained instructions for a talking dragon, the traditional Japanese flapping bird, a traditional frog, and a spider that included several cuts. (Modern origami artists generally shun cutting, but at the age of six I didn't care about such things. Later I felt I redeemed myself by designing a spider that doesn't use cuts.)

Growing up, I kept doing origami, always from books. I devoured origami books, and I haunted libraries and book shops, ever hopeful that a new origami book would appear. Periodically they did: books by Robert Harbin and Samuel L. Randlett in the West and by the Japanese masters Akira Yoshizawa and Isao Honda filled out my early repertoire. I folded everything in each book over and over: birds, fish, mammals, furniture, flowers. I wanted to fold everything in the world from paper, and if I couldn't find something in a book, I tried to make it up.

I started designing my own origami models because I couldn't find examples of the animals I wanted to fold in books—I just had to come up with them on my own. Usually I'd start with an existing model and try to modify it. For example, a goat would become an antelope, and a sparrow mutated into an eagle. At first my "designs" were thinly disguised versions of the models in the books. Over time and with practice, I discovered some new techniques that weren't in the books. What with incorporating these new techniques and getting more adventurous with my variations, I began straying pretty far from the original models in the books. Although there is no definite moment I can point to, eventually I was designing models that were completely different from what was published: I was designing new models.

It's important to point out that I wasn't originally motivated by the goal of doing something different: I just wanted to fold things for which I didn't have instructions. But the thrill of discovering something new is a heady thing, and nowadays I derive a great deal of satisfaction from designing something completely new or from approaching an origami design problem in a way that is different from what's been done before.

The thrill of discovering something new is a hallmark of science, and science in general and mathematics in particular have been a major part of my life. I studied engineering and physics in college and am a laser physicist by profession. Western science and math would seem to be

Ant, Spider, *and* Long-Necked Seed Bug *designed and folded by Robert J. Lang. Photo by Robin Macey.*

very different from an ancient Eastern art like origami, and they've occupied two different parts of my life. But a funny thing happened on the way to the present: Science invaded origami, origami intruded into mathematics, and the result has been some of the most fruitful and fascinating work of my life.

In fact, it's fair to say that origami has been revolutionized by the world of mathematics. In the past decade, mathematicians have begun clarifying the underlying "laws" that define what is possible in origami: how many flaps a model can have (there is no limit—origami sea urchins have hundreds of spines, and origami centipedes have hundreds of legs), what is foldable, and what isn't. The links between origami and other fields are more surprising. For instance, in my own specialty of optics, origami can be used as a tool to study classical lenses, mirrors, and how they redirect beams of light.

The computer offers a new perspective on origami. Several years ago many folders, including me, began to use computers to diagram, or write instructions for, origami designs. This practice was somewhat controversial among some origami artists who felt that using a computer to draw somehow took away some aspect of the art. Recently several folders have begun using the computer to actually design origami figures, applying the mathematical "laws" of origami to very real problems: How do you fold a beetle with six legs? With antennae? With two pairs of wings? With spots on the wings? And so forth. It's fair to say that with the aid of mathematical techniques, origami designers have constructed models that would have been impossible just a few short years ago. For example, I've written a computer program—which I've published on the Internet—that designs the crease pattern for an origami model from a "stick figure" description of the model, based on ideas from origami, mathematics, and even some optics.

With technological progress comes a worry that some folders have occasionally expressed openly: With the progress and use of computers, could a computer one day completely replace the origami designer? I think the answer is no. Despite all the links to mathematics, origami is firmly an art, not a science. Computers, geometry, and math are all tools that origami artists can use. Modern artists may now "paint" with a mouse and a computer screen rather than with a brush and a canvas, but what they produce is still art. So, too, origami artists may wield geometry and computers as tools, but the art will still reside in the folded paper.

ACTIVITY

12

Cube

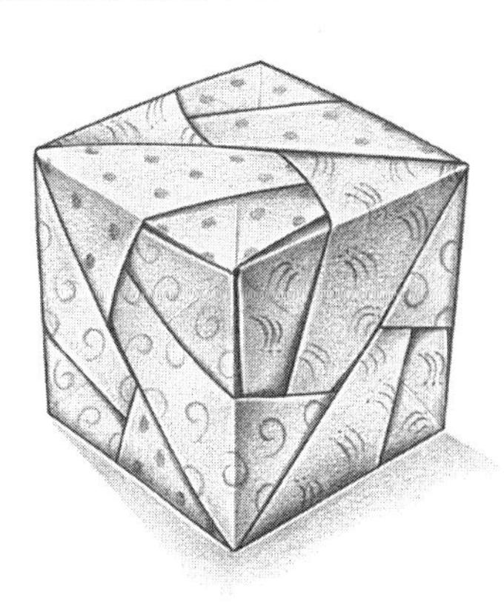

In this activity you will make a cube using six star-building units. As you assemble your cube, try to visualize what the final product will look like. This is easy to do with a cube because it is probably the best-known polyhedron. Also notice the pattern forming on each face of the cube as you tuck the tabs into the pockets.

MATERIALS NEEDED FOR EACH STUDENT

Six 6-inch squares of origami paper—two each of three different colors
Triangular hexahedron from the previous activity
Six 3-inch squares of origami paper—two each of three different colors
Rubber bands
Skewer
Origami journal

GROUPING

Work with a partner. Each person should assemble his or her own cube.

ASSEMBLY INSTRUCTIONS AND QUESTIONS

When you are assembling the cube, think about the answers to the questions asked at each step. When you have finished assembling your model, answer these questions in your origami journal.

Before you start assembling the cube, fold six star-building units, two each of three different colors. Record the colors you used in the rectangles at the right.

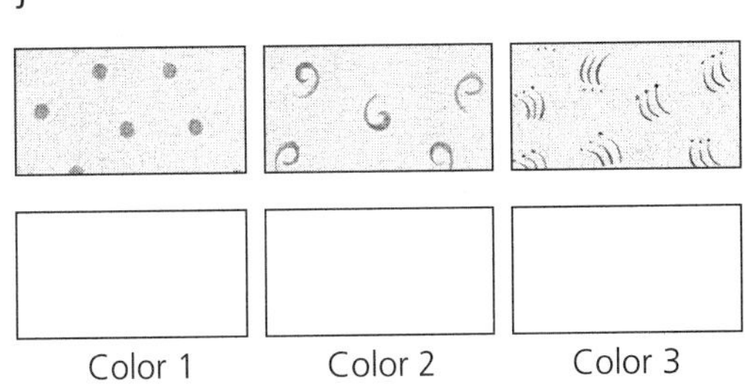

1 Fit a tab of a color 2 unit into a pocket of a color 1 unit as shown.

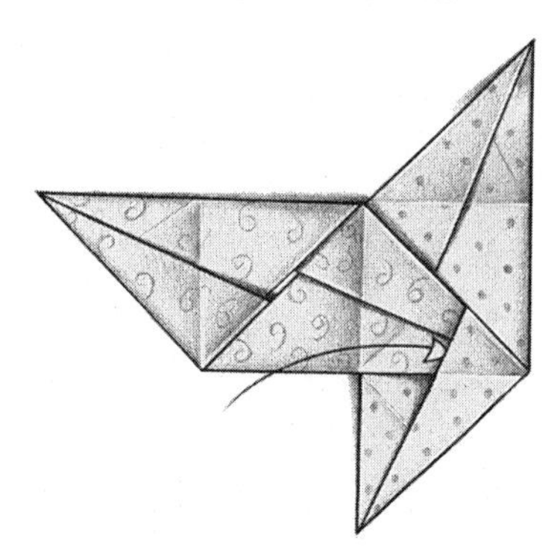

2 Rotate the figure 90° counterclockwise. Next, fit a tab of the color 3 unit into the pocket of the color 2 unit as shown.

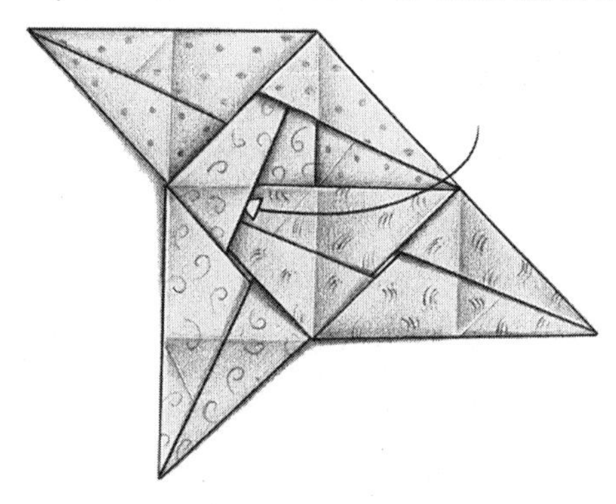

3 Form a pyramid by inserting the color 1 unit into the color 3 unit. You have now formed one corner of the cube. Notice that the pockets are on the outside of the cube.

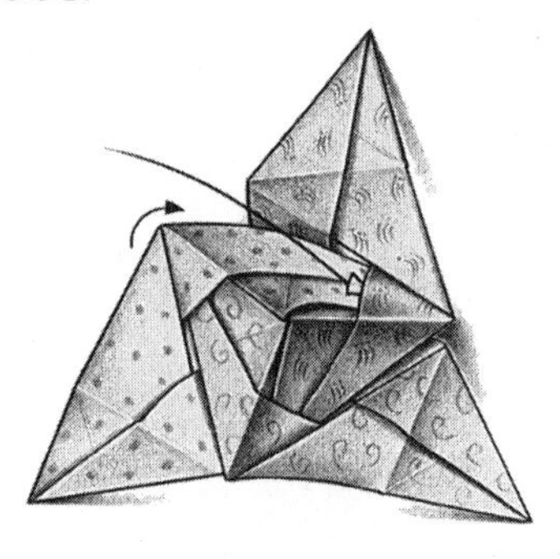

What kind of pyramid is formed in this step?

4 Complete the cube with the three remaining star-building units. The following picture shows three faces of the completed cube. Make sure that each face has only two colors on it.

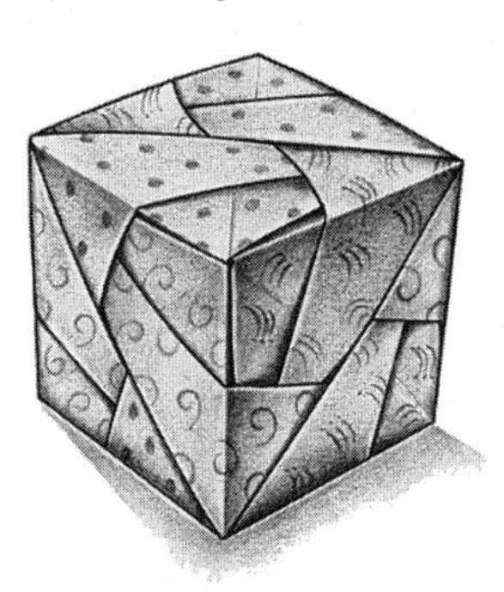

Each square face is composed of only two colors. Opposite faces have the same colors. How can these facts help you decide what color to add to complete each face?

5 You have made a cube. Notice that there is a one-to-one correspondence between tabs and pockets. There are no stray tabs and no empty pockets. If you have empty pockets, check inside the cube to see if there are any stray tabs.

Remember to answer each of the assembly questions in your origami journal.

EXPLORING YOUR MODEL

Record your answers to the exploring questions in your origami journal.

1. Use rubber bands to outline the reflection planes of the cube. Don't forget that a plane can cut a cube diagonally. Draw a sketch to show each reflection plane.
2. Use skewers to find the axes of rotational symmetry of the cube (you may have to puncture the paper with the skewers). How many axes of 4-fold rotational symmetry are there? How many axes of 3-fold rotational symmetry are there? How many axes of 2-fold rotational symmetry are there? (For the axes of 2-fold symmetry, it may be difficult to use skewers. Rotate the cube between two fingers instead.)
3. Describe the rotational symmetries of the cube using degrees.
4. Assemble a second cube. This time, use square papers whose sides are half the length of your original paper. Copy the chart below into your origami journal. Make predictions in the bottom row of the chart. Then fill in the rest of the chart. How accurate were your predictions?

COMPARING TWO CUBES

	Edge length of paper	Edge length of cube	Surface area of cube	Volume of cube
Original cube				
Smaller cube				
Ratio: $\frac{\text{Original cube}}{\text{Smaller cube}}$				

5. Find the *exact* length of a side of the cube. (Hint: Unfold one star-building unit, look at the fold lines, and find the outline of one of the cube faces in the fold lines.)
6. Look at the colors on your cube. Describe one of the patterns that you see.

7. How many different-looking cubes can you make from your six star-building units? Explain your reasoning.

8. Try building different-sized cubes. (You might want to see who can build the smallest one and who can build the largest one.) Create a chart similar to the one below. Collect results from your group or your class.

Name of student	Edge length of paper (in.)	Edge length of cube (in.)	Surface area of cube (sq in.)	Volume of cube (cu in.)

9. Make three separate graphs to show the relationship between the edge length of the paper and the edge length of the cube, the edge length of the paper and the surface area of the cube, and the edge length of the paper and the volume of the cube. Describe how the graphs differ.

Cube

OBJECTIVES

To assemble a cube using star-building units
To identify patterns in the assembly of a cube
To discover the reflection and rotational symmetries of a cube
To compare the area of the faces and volume of two cubes whose edges are in a ratio of 1:2

MATERIALS NEEDED FOR EACH STUDENT

Six 6-inch squares of origami or patty paper—two each of three different colors
Triangular hexahedron from the previous activity
Six 3-inch squares of origami or patty paper—two each of three different colors
Student worksheet pages—one copy for each pair of students
Rubber bands
Skewer
Origami journal

TEACHER MATERIALS

Three large squares folded into three star-building units for demonstration
Completed cube

TIME

One class period

GROUPING

Students should work with a partner. Each student will assemble his or her own cube.

GENERAL INSTRUCTIONS

Students should have little difficulty assembling the cube from the six star-building units. Be sure to emphasize that color arrangement in the final model is important.

A possible extension is to have students fold and assemble the open-frame polyhedra found on pages 62–67 of *Unit Origami* by Tomoko Fusè.

ANSWERS

Folding Questions

1–2. [No questions.]

3. The pyramid formed is a triangular pyramid with a triangular base.
4. Answers will vary. You can look at the two colors of an incomplete face and decide which of the two colors belongs in the empty pocket. Once one face is complete, you know the opposite face will be the same.
5. [No question.]

Exploring Your Model

1. The cube has many symmetries. Shown in the figure at right are three planes of reflection symmetry that are fairly easy to find. Either you or the students could also demonstrate the planes of reflection symmetry using a large piece of paper or a plastic transparency. Cut a square hole the size of the face of the cube in the paper or the transparency and slide the cube halfway through the paper.

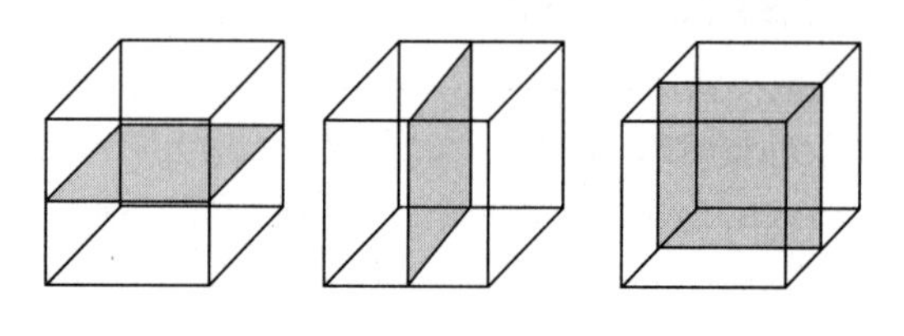

The six reflection planes that cut through the cube diagonally (connecting pairs of opposite edges) are less obvious. Two of these planes are shown at right. Again, you can show the reflection plane by cutting a rectangular hole in paper or a transparency.

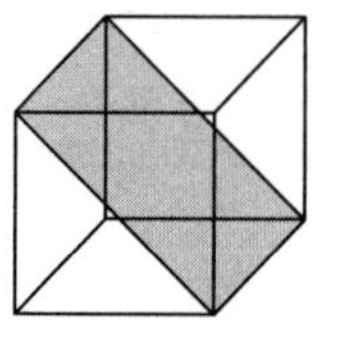

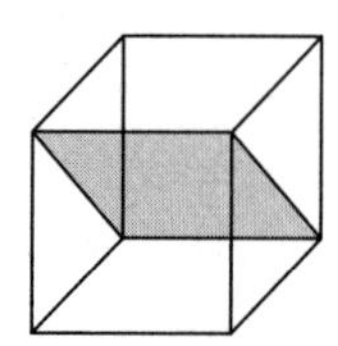

2. Skewers can help students see the axes of rotational symmetry.

There are four axes of 2-fold rotational symmetry.

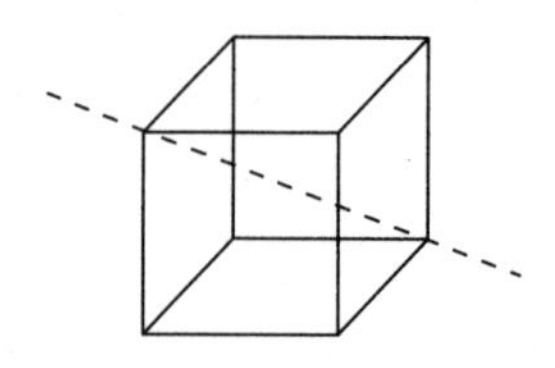

There are three axes of 4-fold rotational symmetry.

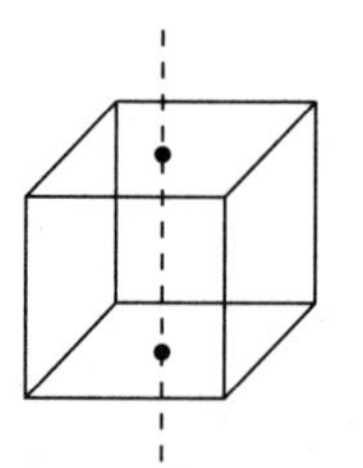

There are six axes of 2-fold rotational symmetry. There are too many layers of paper to puncture them easily with a skewer. Students will have to rotate the figure between their fingers instead. Each axis passes through the midpoints of diagonally opposite edges.

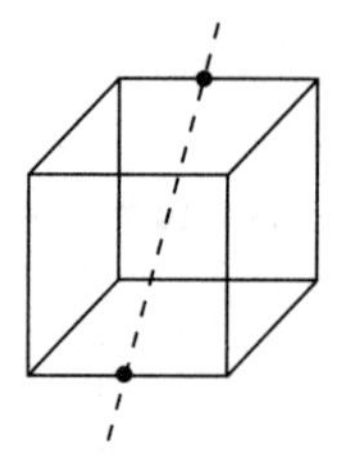

3. Four-fold rotational symmetry is also called 90° rotational symmetry. (Two-fold, or 180° rotational symmetry, is sometimes called *point symmetry.*)
4. Answers will vary depending on the size of the original square.

COMPARING TWO CUBES

	Edge length of paper	**Edge length of cube**	**Surface area of cube**	**Volume of cube**
Original cube	6	2	24	8
Smaller cube	3	1	6	1
Ratio: $\frac{\text{Original cube}}{\text{Smaller cube}}$	$\frac{6}{3} = 2$	$\frac{2}{1} = 2$	$\frac{24}{6} = 4$	$\frac{8}{1} = 8$

5. Answers will vary depending on the size of the original square.

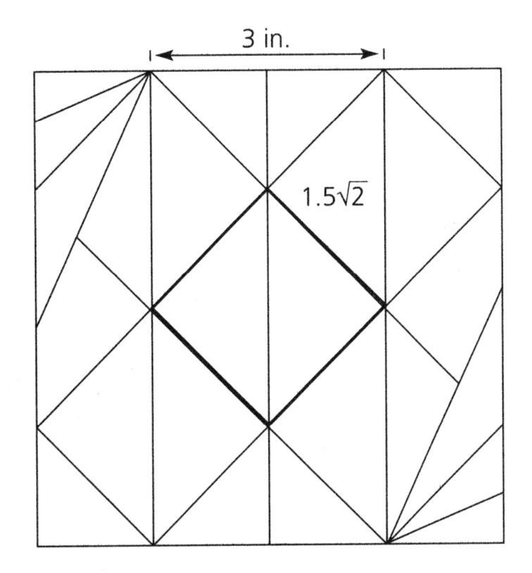

6. Answers will vary. Some students may notice that the colors appear as "bands" that go around the cube in three different directions.
7. You can build six different cubes if you start with three colors and two star-building units folded from each color. The reason for this is that you have three choices for the first color you choose, and then you have two choices for the next color, then one choice for the last color. $3 \times 2 \times 1 = 6$ choices.
8. Answers will vary.
9. The graph of edge length of the paper versus edge length of the cube will be linear. The graph of edge length of the paper versus surface area of the cube will be parabolic. The graph of edge length of the paper versus volume of the cube will be cubic.

ACTIVITY

13

Two Colliding Cubes

A pattern is forming. A triangular hexahedron is made of three star-building units, and a cube is made of six units. You guessed it—in this activity, you will be assembling a model made with nine star-building units. You will create two colliding cubes.

Your work with the simple cube will help you with this polyhedron. Studying the picture at the right to see how one corner of a cube is assembled will also help you with the colliding cubes model. Count the colors that make up a corner of one of the cubes.

MATERIALS NEEDED FOR EACH STUDENT OR PAIR OF STUDENTS

Nine squares of origami or patty paper for nine star-building units—
three each of three different colors
Rubber bands
Skewers
Masking tape
Origami journal

GROUPING

Work with a partner. You can each make your own colliding cube, or you can make one together.

SSEMBLY INSTRUCTIONS

1 Fold nine star-building units, three each of three different colors. Make a valley fold on each unit as shown.

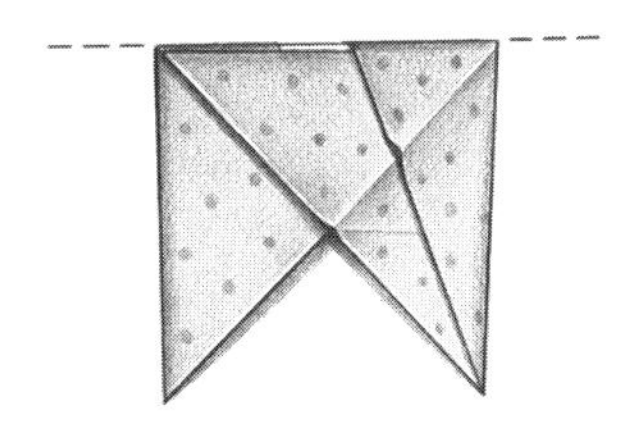

2 Assemble a cube using the method from the previous activity.

3 Deconstruct one of the pyramids at one of the vertices of the cube by pulling three tabs out of three pockets. Now there are three loose tabs.

4 Choose one of the loose tabs. Attach two new units to the loose tabs to make a new pyramid.

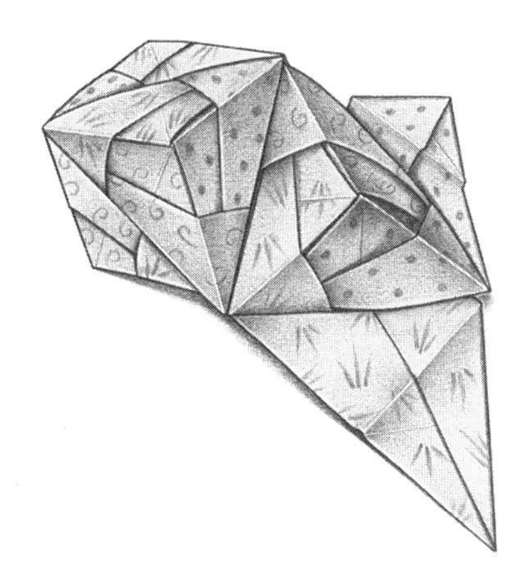

5 A new cube is beginning to form. Complete the cube by adding the last unit.

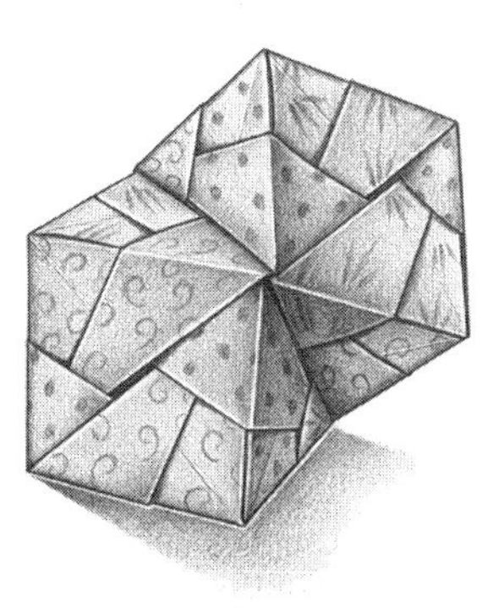

EXPLORING YOUR MODEL

Record your answers to the exploring questions in your origami journal.

1. Why do you think this model is referred to as two colliding cubes rather than simply double cubes?

2. For a convex polygon, any diagonal, a line segment connecting two nonconsecutive vertices, lies on or inside the polygon. For a nonconvex polygon, at least one diagonal lies outside the polygon. Which of the polygons below is convex and which is nonconvex?

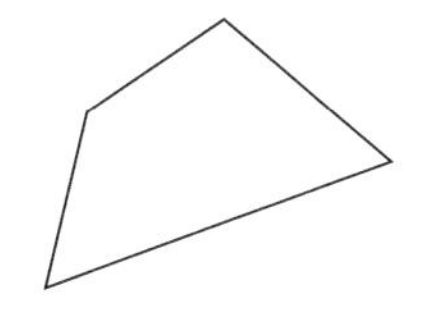

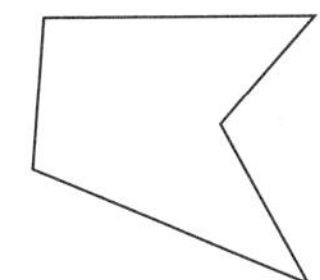

3. Now think about three dimensions. Is a cube convex? Is your colliding cube model convex? Explain. Write descriptions of a convex and a nonconvex polyhedron.

4. Count the number of edges, faces, and vertices on your colliding cubes. Make a chart, or use one provided by your teacher, to record the number of edges, faces, and vertices for the triangular hexahedron, the cube, and the colliding cubes. Find a formula that relates the number of edges, faces, and vertices.

5. The colliding cubes model is not a regular polyhedron. In a regular polyhedron, all faces are congruent, and the same number of edges meet at each vertex. Which parts of the definition of a regular solid do not hold true for the colliding cubes?

6. a. If the area of one square face of the colliding cubes is 4 square units, what is the total surface area?

   b. If the area of one square face of the colliding cubes is 9 square units, what is the total surface area?

   c. If the area of one square face of the colliding cubes is x square units, what is the total surface area?

7. What fraction of the volume of each cube is hidden from view?

8. Find the planes of reflection symmetry and the axes of rotational symmetry of the colliding cubes model. You might want to use rubber bands and skewers, or you could hang the model to figure out the answers.

Two Colliding Cubes

OBJECTIVES

To assemble the colliding cubes
To define convex and nonconvex polyhedra
To discover Euler's formula relating edges, faces, and vertices
To explore the surface area of the colliding cubes
To find the reflection and rotational symmetry of the colliding cubes

MATERIALS NEEDED FOR EACH STUDENT OR PAIR OF STUDENTS

Nine squares of origami paper for nine star-building units—three each of three different colors
Rubber bands
Skewers
Masking tape
Student worksheet on Euler's formula—one copy for each pair of students
Origami journal

TEACHER MATERIALS

Nine large star-building units for demonstration
Completed model of two colliding cubes

TIME

One class period

GROUPING

Students work in groups of four, with a designated partner. Each person can make his or her own colliding cube model, or two students can make one together.

GENERAL INSTRUCTIONS

The written instructions and diagrams for assembling the colliding cubes may be difficult for some students to follow. It will be helpful to have a completed model available for them to examine.

As an extension you could have students fold and assemble the pinwheel colliding cubes model pictured on page 137 of *Unit Origami* by Tomoko Fusè.

ANSWERS

Exploring Your Model

1. Answers will vary. The model shows intersecting cubes. The cubes themselves are not complete cubes.
2. The first polygon is convex and the second is nonconvex.
3. A cube is convex. The colliding cube model is a nonconvex polyhedron. There are many ways to describe convex and nonconvex polyhedra. The following are some possibilities:
   - Any line segment connecting two points on the surface of a convex polyhedron lies on or inside the polyhedron. At least one line segment connecting two points on the surface of a nonconvex polyhedron lies outside the polyhedron.
   - A convex polyhedron rests entirely on the plane of each of its faces. This does not hold true for a nonconvex polyhedron.
   - If a convex model were shrink-wrapped, there would be no air pockets; that is, the object is shrink-wrapable. This does not hold true for a nonconvex polyhedron.
   - Every dihedral angle formed by two faces meeting at an edge of a convex polyhedron is less than 180°. This is not so for a nonconvex polyhedron. (A dihedral angle is the angle between two faces on the inside of the polyhedron.)

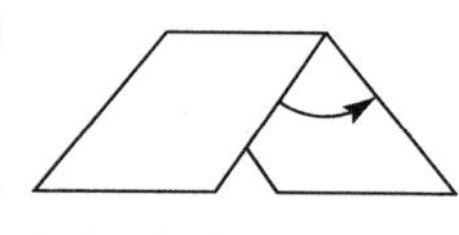

Dihedral angle

4. The colliding cubes model has 12 faces, 11 vertices, and 21 edges. (Students may want to use small pieces of masking tape to count the edges, faces, and vertices on their colliding cubes model.) Euler's formula is as follows:
   The number of faces + the number of vertices = the number of edges + 2.
   Euler's formula works for the colliding cubes model since 12 + 11 = 21 + 2.

EULER'S FORMULA

Polygon	Faces	Vertices	Faces + vertices	Edges	Does Euler's formula work?
Triangular hexahedron	6	5	11	9	yes
Cube	6	8	14	12	yes
Two colliding cubes	12	11	23	21	yes

5. The colliding cubes model does not have congruent faces; six of its faces are not regular polygons (they are right isosceles triangles), and it does not look the same at each vertex. At some vertices, three faces and three edges meet. At others, six faces and six edges meet.
6. a. If one square face is 4 square units, then the surface area is 36 square units.
   b. If one square face is 9 square units, then the surface area is 81 square units.
   c. If one square face is x square units, then the surface area is $9x$ square units.
7. One-fourth of the volume of each cube is hidden from view.
8. The colliding cubes model has four planes of reflection symmetry, one axis of 3-fold rotational symmetry, and three axes of 2-fold rotational symmetry. Two views of the colliding cubes model are shown below.

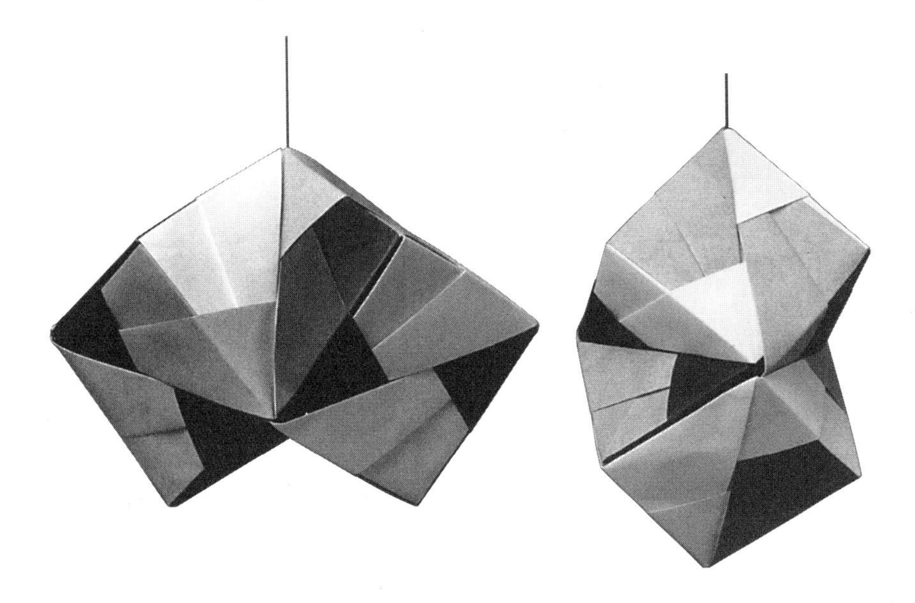

Euler's Formula

Polygon	Faces	Vertices	Faces + vertices	Edges	Does Euler's formula work?
Triangular hexahedron	6	5	11	9	yes
Cube	6	8	14	12	yes
Two colliding cubes	12	11	23	21	yes

ACTIVITY 14

Stellated Octahedron

Assembling a stellated octahedron can give you a satisfying feeling of accomplishment. *Stellated* means "star-shaped," so you can see how this polyhedron got its name. The picture of the stellated octahedron makes it look difficult to assemble, but it isn't. The procedure for making each point or pyramid of the stellated octahedron is the same as that for all the other models you have made using star-building units. It is the pyramid that gives this model its name and its stability.

MATERIALS NEEDED FOR EACH STUDENT OR PAIR OF STUDENTS

Twelve squares of origami paper for twelve star-building units—three each of four different colors
Skewers
Masking tape
Origami journal

GROUPING

Work with a partner. You can each make your own stellated octahedron, or you can make one together.

ASSEMBLY INSTRUCTIONS AND QUESTIONS

Record your answers to the assembly questions in your origami journal.

Before you start assembling the stellated octahedron, you will need to make twelve star-building units in four different colors or patterns. You might want to color the rectangles at the right so that you can keep track of the color numbers as you follow the assembly directions.

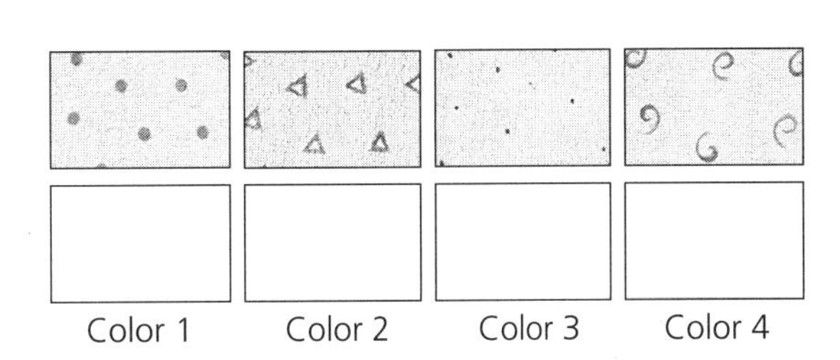

1 Make a valley fold on each unit and unfold, as shown.

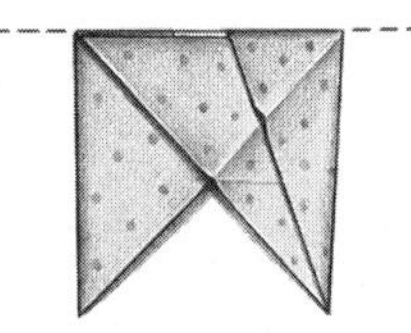

2 Select three units, each of a different color. Assemble a three-color pyramid as you have done for the previous models.

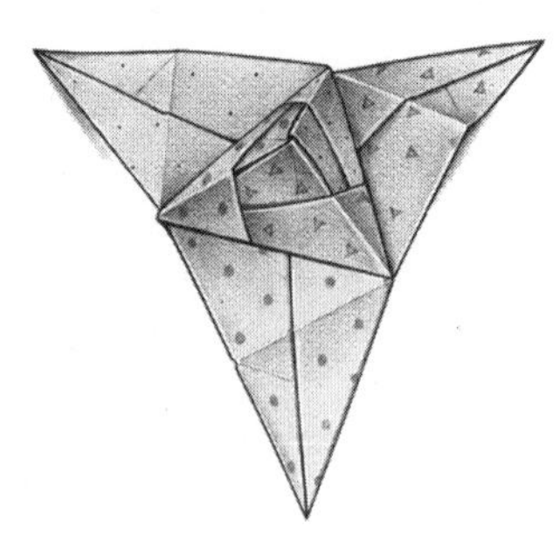

3 Start to build another pyramid by choosing a color you already have. Follow this rule: when you insert the unit into a pocket, each valley (or bent square) should be composed of only two colors.

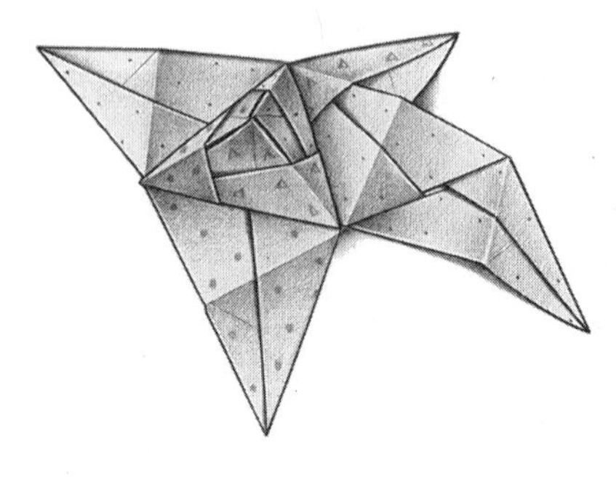

4 To complete the second three-color pyramid, use a new color, color 4.

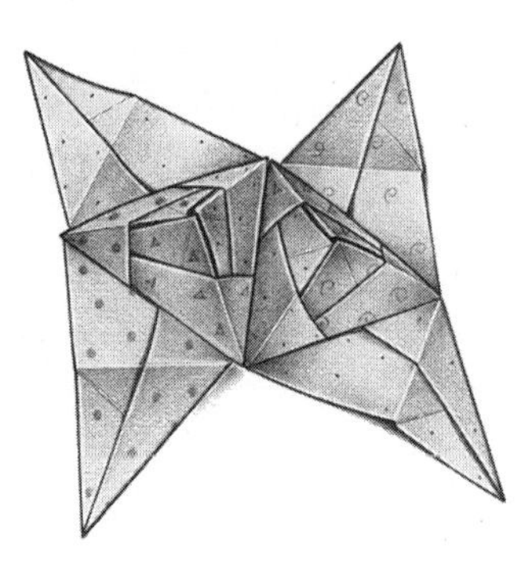

5 Continue in a clockwise direction. Start your third pyramid by matching the color pattern in the valley, or bent square. Finish it with a unit that is the same color as the one at the start of the circuit of pyramids.

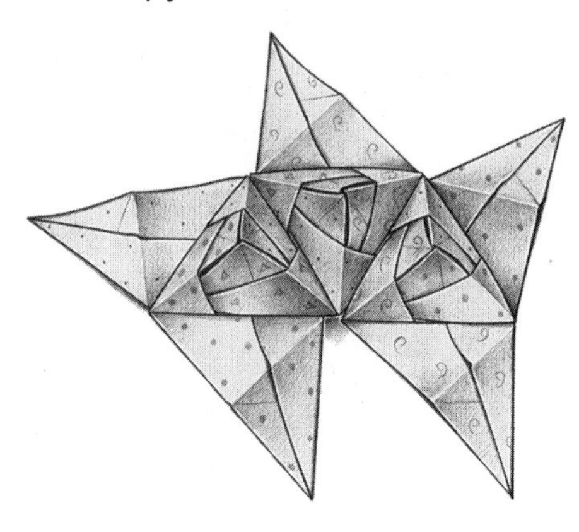

6 You will need to add one last unit to make the fourth pyramid, which completes the circuit. Follow the rule of matching colors on both sides of a valley, or bent square.

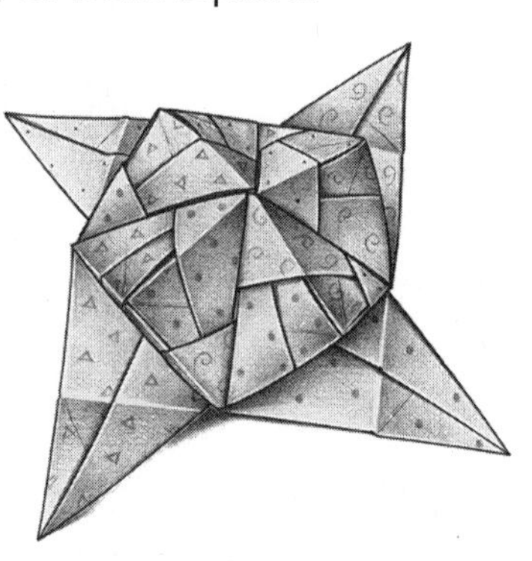

When you look down on the four pyramids you have formed, what shape do you see?

7 Turn the model over so that it looks like a bowl. Repeat the process of forming pyramids. Start with any loose tab. Form a pyramid with two new units. The color of the first unit will be determined by the existing colors in the valley, or bent square. The color of the second unit should be the same as the color of the tab to its right.

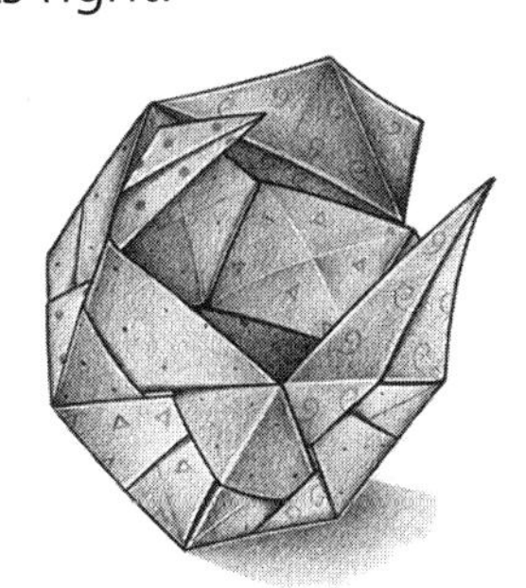

8 Continue the process with the remaining units until you have a stellated octahedron.

EXPLORING YOUR MODEL

Record your answers to the exploring questions in your origami journal.

1. How many pyramids are there in the completed model of the stellated octahedron? If you cut off each pyramid of the stellated octahedron, what would the resulting polyhedron be? What shape would each face of this core polyhedron be?
2. What is the length of each edge of the polyhedron at the core of the stellated octahedron?
3. What is the surface area of the core polyhedron?
4. Describe the stellated octahedron in terms of the number of pyramids it has and how many edges meet at each vertex.
5. Pick one color you used for your stellated octahedron. Find it on the polyhedron you made. Can you see how that color forms a band around the polyhedron? Do you think the same thing happens with the other colors you used? Find out.
6. Euler's formula works for the stellated octahedron. Count the number of faces and vertices and use Euler's formula to find the number of edges. Add the stellated octahedron to your Euler's formula chart.

7. Find the planes of reflection symmetry and the axes of rotational symmetry of the stellated octahedron. You might want to use rubber bands and skewers, or you can hang your model to figure out the answers.

8. You used twelve star-building units to assemble a stellated octahedron. To figure out how many star-building units it would take to assemble a stellated icosahedron, look at the following chart:

	Number of stellated faces	**Star-building units needed**
Stellated tetrahedron	4	3
Stellated hexahedron	6	6
Stellated octahedron	8	12
Stellated icosahedron	20	?

Stellated Octahedron

OBJECTIVES

To assemble the stellated octahedron
To find the dimensions and surface area of the core of the stellated octahedron
To use Euler's formula to find the number of edges on the stellated octahedron
To predict the number of units needed for a stellated icosahedron

MATERIALS NEEDED FOR EACH STUDENT OR PAIR OF STUDENTS

Twelve squares of origami or patty paper for twelve star-building units—three each of four different colors
Skewers
Masking tape
Student worksheet pages—one copy for each pair of students
Origami journal

TEACHER MATERIALS

Twelve large squares of paper folded into star-building units for demonstration
Completed stellated octahedron

TIME

One to two class periods

GROUPING

Students should work in groups of four, with a designated partner. Each student can make his or her own stellated octahedron, or two partners can make one together.

GENERAL INSTRUCTIONS

Students may have difficulty following the written assembly directions. If possible, have a completed model available for them to examine. Be sure the color arrangement in students' completed models is correct. Since students know how to fold the star building units, you might ask them to fold the units as homework the night before doing this activity in class.

Making true stellation of a polyhedron involves extending the faces of the polyhedron until the faces intersect. The true stellation of an octahedron results in a stella octangula.

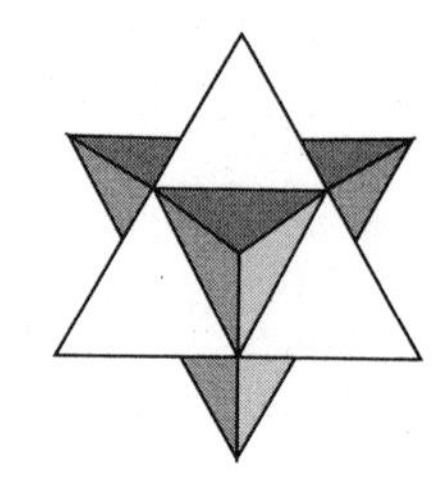

David Masunaga's students have made stellated octahedron earrings using $\frac{3}{4}$-inch squares. Some of your students may be interested in this unusual project.

ANSWERS

Assembly Questions

1–5. [No questions.]

6. You will see an octagon.

7–8. [No questions.]

Exploring Your Model

1. There are eight pyramids. Each pyramid would be replaced by an equilateral triangle. An octahedron is at the core of the stellated octahedron.
2. The length of each edge of the core octahedron is the same as the length of each edge of the stellated octahedron. (Using 6 in. squares, the length is 3 in.)
3. Using 6 in. squares, each edge of the core octahedron is an equilateral triangle measuring approximately 3 in. on a side. Using the Pythagorean theorem, you can determine the height of the triangle: $\sqrt{6.75} \approx 2.6$. The area of the triangle is $\frac{1}{2} \times 3 \times 2.6 \approx 3.9$ sq in. There are eight faces, so the surface area is about 31.2 sq in.

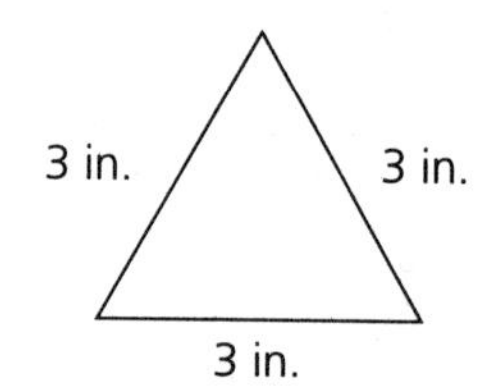

4. Answers will vary. Here is one possibility: The figure has fourteen vertices. There are eight triangular pyramids. At each of six vertices, eight edges meet. At eight vertices (the "peaks" of the pyramids), only three edges meet.
5. Answers will vary. Students may notice that all colors form bands around the stellated octahedron. Some may even notice that opposite pyramids have the same colors, but that the colors rotate in opposite directions. Color symmetry will be introduced in Activity 15.

6. There are fourteen vertices and twenty-four faces. Students' strategies for counting the faces and vertices will vary. Some may use patterns they see on the polyhedron. Others may use small pieces of masking tape to keep track of their counting on the polyhedron. Using Euler's formula, they will get: $24 + 14 = E + 2$. Therefore, the number of edges is 36.

Polygon	**Faces**	**Vertices**	**Faces + vertices**	**Edges**	**Does Euler's formula work?**
Stellated octahedron	24	14	38	36	yes

7. The stellated octahedron has twelve planes of reflection symmetry. It also has 2-fold, 3-fold, and 4-fold rotational symmetry. It has a total of thirteen axes of rotational symmetry. Have students describe the axes of rotational symmetry of the stellated octahedron. Skewers can be helpful. You may want to hang the model as well. There are three ways to hang this model.
   - One way shows there is a 4-fold rotational axis through every pair of vertices at which eight edges meet. There are three such axes.
   - Another way shows there is a 3-fold rotational axis through vertices of pairs of opposite pyramids. There are four such axes.

   - A third way shows there is a 2-fold rotational axis through the midpoints of pairs of edges formed by two pyramids. There are six such axes.

8. Students can extend the pattern on the chart to find that the stellated icosahedron needs 30 star-building units. Some students may use a proportion, such as the following one, which involves a stellated octahedron and a stellated icosahedron:

   $\frac{8}{12} = \frac{20}{x}$, where x is the number of units needed for a stellated icosahedron.
   So, $x = 30$.

ACTIVITY 15

Stellated Icosahedron

In this activity you will build a very beautiful model of a stellated icosahedron. One of the reasons that this model is so striking is that the colors are arranged symmetrically. You will explore the possible color combinations and discover that there are several color arrangements you can use to make "different" stellated icosahedra. Even though the model looks difficult to assemble, you will find that it is quite easy if you follow the two basic rules given below.

MATERIALS NEEDED FOR EACH PAIR OR GROUP OF STUDENTS

30 star-building units—five each of six different colors
Origami journal

GROUPING

Work with a partner or a group of four. Each pair or group can make a stellated icosahedron.

ASSEMBLY INSTRUCTIONS

When assembling this model, follow these two rules:

Rule 1: Always make sure that three colors meet at the vertex of each star or pyramid.

Rule 2: For each valley (or bent square) between pyramids, be sure the same color combination is on both sides of the valley.

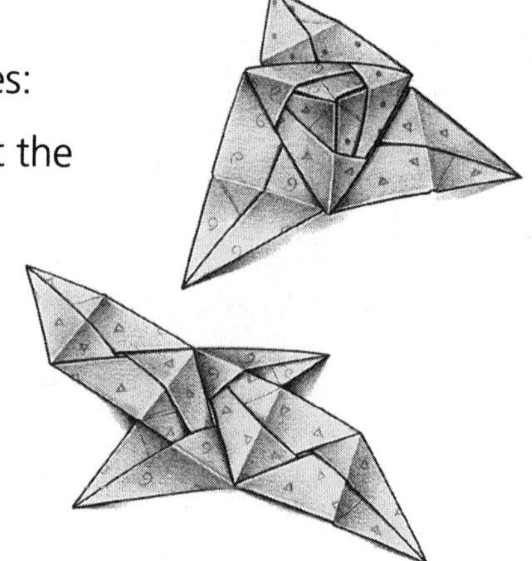

efore starting to assemble the stellated icosahedron, you'll need to fold hirty star-building units. To help you keep track of the colors for this model, olor the rectangles at the right. ut all star-building units of one olor (color 6) aside. Group the ther star-building units by color.

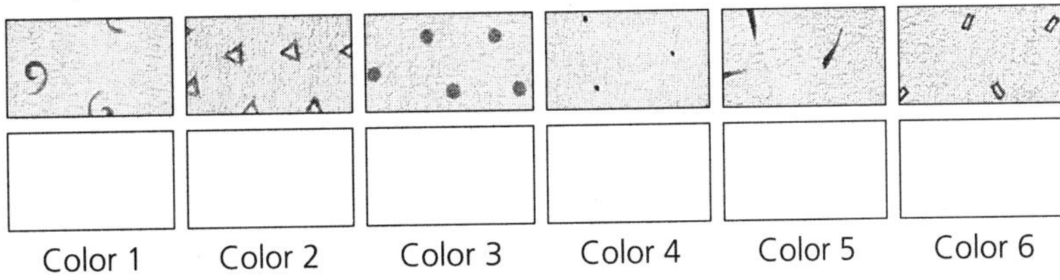

1 Make one pyramid using colors 1, 2, and 3.

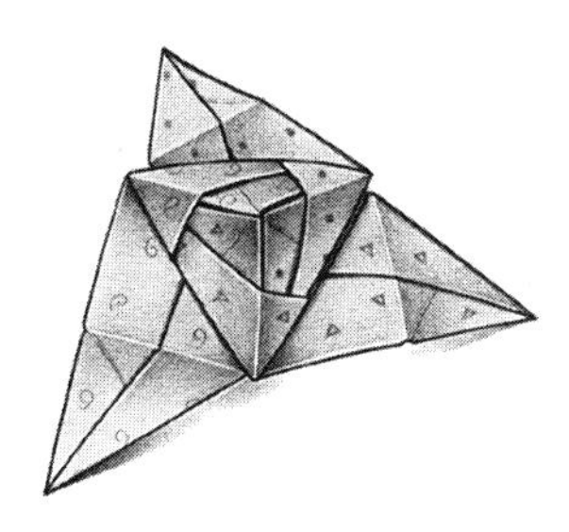

2 Find a valley, or bent square. Insert the color needed to make this bent square the same colors on both sides of the valley. Now you have the start of the second pyramid.

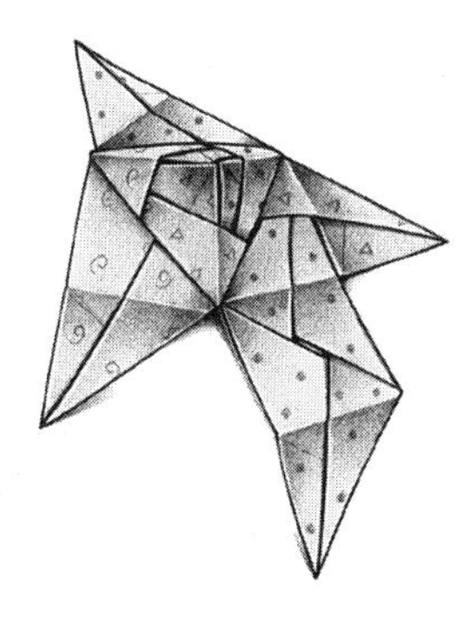

3 Complete the second pyramid with color 4. (You will make five pyramids meeting at each vertex.)

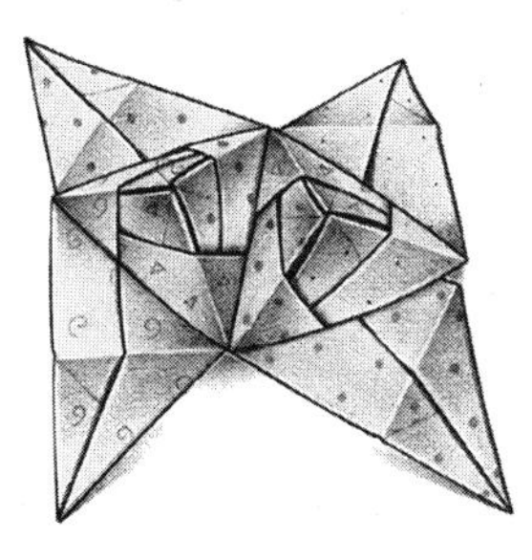

4 Match the color in the bent squares, and use a color 5 unit to complete the third pyramid.

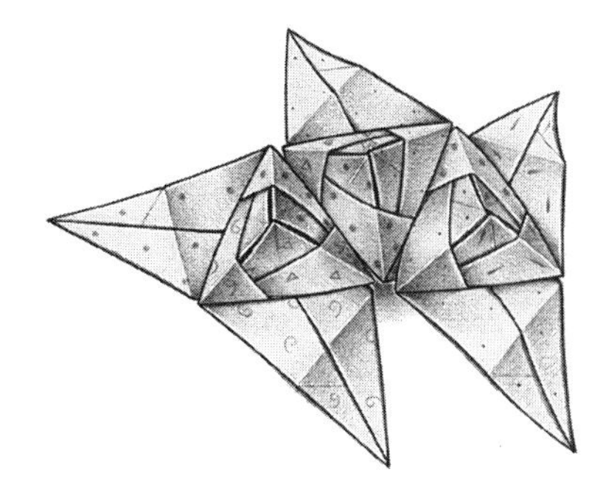

5 Complete the last two pyramids using the rules of three colors at a vertex and the same colors on both sides of a bent square.

6 Now you will be attaching a belt of ten pyramids to the "bowl" of five pyramids you already made. To start, find one incomplete bent square. Use the rules to decide which color to insert to complete the bent square. Complete the pyramid you are working on by using color 6.

7 Continue working around the model following the same rules. The rules will guide you. Use color 6 in each of the ten pyramids of the belt.

8 Build another group of five pyramids meeting at a vertex. Build them, one at a time, onto the belt of pyramids you just completed. Continue following the rules.

9 You have made a stellated iscosahedron

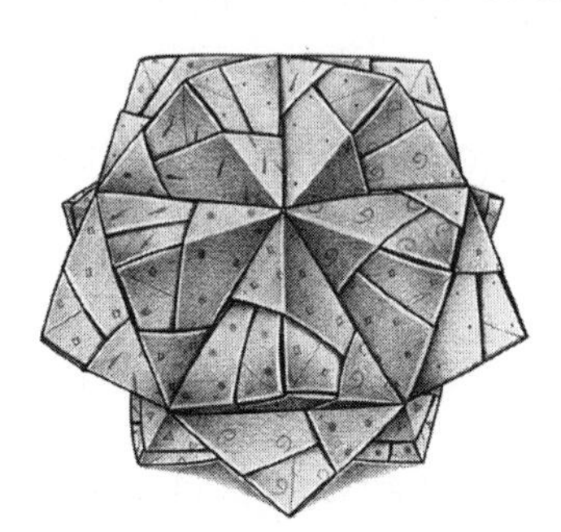

EXPLORING YOUR MODEL

Record your answers to the exploring questions in your origami journal.

1. You consciously built a band of color 6 around the middle of your model. Look for other bands of color. How many can you find? How would you describe what happened when you followed the pattern of putting the same colors on both sides of every bent square?

2. Pick a vertex on one of the pyramids. Name the three colors that meet at that vertex. Look at the vertex on the other side of the model that is directly opposite the one you picked. What colors meet at this second vertex? Are the colors arranged in exactly the same way for both vertices? Explain.

3. For any pair of opposite vertices, will you find the same three colors? Explain why this happens by looking at one particular vertex again. (Hint: Think about the bands of color.)

4. You used six different colors to make your model. There are three different colors on each pyramid. How many different ways are there of choosing three different colors at a time, if you have six colors to choose from? Another way to ask the same question is, How many distinctly colored pyramids can you make using six different colors? Explain your strategy for finding the solution. Were you able to use a pattern to help you?

5. How many faces are there on an icosahedron? How many pyramids on a stellated icosahedron? Do all color combinations appear on the model you built? Explain your conclusions.

6. Using your results from question 5, determine how many distinctly colored models of the stellated icosahedron you can make using the same six colors.

7. Pretend you are an artist investigating all the possible combinations of six colors: three primary colors (red, yellow, and blue) and three secondary colors (orange, purple, and green). As you discovered in question 4, if you begin with six colors and choose three colors at a time, there are twenty possible combinations. Although there are twenty pyramids on the stellated icosahedron, because opposite pyramids have the same color combinations, you can use only ten of the possible color combinations when constructing the model. You can create an artist's set of color combinations by constructing a pair of stellated icosahedra that show all of the possible combinations of six colors chosen three at a time. Determine a strategy for doing this, and build the two models.

Stellated Icosahedron

OBJECTIVES

To build a stellated icosahedron with specific color restrictions
To describe color symmetries
To find combinations of colors
To make an organized list of color combinations

MATERIALS NEEDED FOR EACH PAIR OR GROUP OF STUDENTS

30 star-building units—five each of six different colors
Student worksheet pages—one copy for each pair of students
Origami journal

TEACHER MATERIALS

A completed one-color model of a stellated icosahedron (optional)

TIME

One or two class periods

GROUPING

Students should work in pairs or groups of four. Each pair or group will make a stellated icosahedron.

GENERAL INSTRUCTIONS

The level of difficulty of this activity will depend on how many activities students have done previously using the star-building units. The assembly of this model is surprisingly easy if you follow the two basic rules. If this activity is done with a group, the actual construction of the model will not take very long. You will get dramatic results if some groups use six-inch paper while other groups use three-inch paper. In the artist's palette activity described in question 7, students create a beautiful complementary pair of models.

When students look at their models, they will notice that the bands of color weave in and out. To explore this further, have students study the strip modular model of a stellated icosahedron in the article "Some Isonemal Fabrics on Polyhedral Surfaces" by Jean Pedersen in the book *The Geometric Vein.*

ANSWERS

Exploring Your Model

1. There are six bands of color, one for each color used. Each band goes around the model from one pole to the opposite pole and back.
2. Each pair of vertices on opposite sides of the model have the same color combinations. If students look at the vertices carefully, they will notice the color arrangements are mirror images of each other.

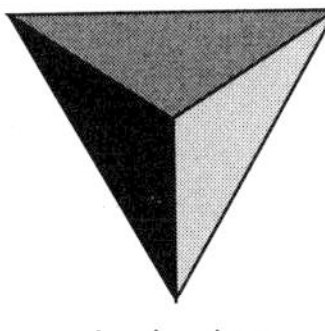

clockwise

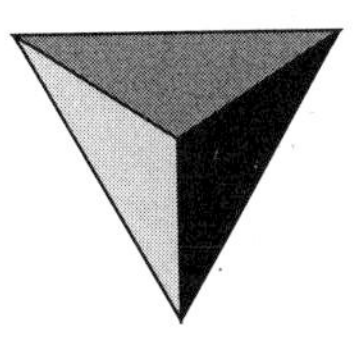

counterclockwise

3. Answers will vary. One possibility is to have students think of the top of the model (where their chosen vertex is located) as the north pole and the bottom of the model as the south pole. If a band of color passes through the north pole, it will also pass through the south pole. Thus, the same three colors will meet at opposite poles.
4. There are twenty distinct ways of choosing three different colors, given a total of six colors. They are shown in the following organized list.

RBY	RYP	ROG	BYG	YOP
RBO	RYG	RPG	BOP	YOG
RBP	ROY	BYO	BOG	YPG
RBG	ROP	BYP	BPG	OPG

5. There are twenty faces on an icosahedron and twenty pyramids on a stellated icosahedron. In exploration question 2, students discovered that the pyramids come in pairs, with each pair having the same color combination. Therefore, only ten of the twenty color combinations are represented on each stellated icosahedron.
6. There are twenty possible color combinations, and ten color combinations are used to make a stellated icosahedron. Therefore, ${}_{20}C_{10} = 184{,}756$ distinctly colored stellated icosahedrons that can be made using only six colors. You might ask students what the probability is that any two of them would make exactly the same model.
7. Students construct two models of the stellated icosahedron.

My Experiences with Origami

Laura Kruskal

Laura Kruskal

Laura Kruskal, a resident of Princeton, New Jersey, is an original creator of origami models and a renowned teacher whose approach is to make origami accessible to all. To Laura, origami is an international language, unifying people from all different backgrounds and cultures. She has won several awards, and her models have been frequently published in books and magazines all over the world.

I have had so many wonderful origami experiences. My husband's job as a professor of mathematics takes us all over the world, and it has been on these trips that I've had some of my most memorable experiences.

In April 1989, my husband and I traveled to China in response to his invitation from the Chinese Academy of Sciences to lecture and consult at their most prestigious universities and scientific institutes. Thus, we were in Beijing on May 15 when Soviet president Mikhail Gorbachev arrived for the first Sino-Soviet summit in thirty years.

At the time of President Gorbachev's visit, I was teaching English at Beijing University, using origami as a tool for increasing vocabulary and conversational skills. The students in my classes, emboldened by the presence of Gorbachev—the father of *glasnost*—and a large contingent of the Western press, joined the students in Tiananmen Square, who were on their third day of a hunger strike. The protest was so peaceful that I was able to go to the square every day and continue folding with the students.

One day I reminded my students that it was the Chinese who invented paper in the first century A.D. and subsequently the art of paper folding, so surely someone in the class could teach me an origami model. One of the students folded and presented me with an action model her grandmother had taught her to fold. I brought the model home with me, to discover that it had never before been seen in the West. I call it *Drinking Bird*. It has become one of the most popular models in my folding repertoire.

One summer, when my husband was attending an international conference in Rome, I was invited to the British Embassy to participate in a birthday celebration for Queen Elizabeth. Because I had nothing elegant enough to wear, I borrowed a dress from a friend. Not having matching accessories, I located a cartoleria (paper shop) and found an exquisite gift wrap, which I folded into a handbag, a wallet, a change purse, and a tissue holder.

At the birthday celebration, while I was in the reception line shaking hands with the ambassador's wife, she looked admiringly at my handbag and commented on how perfectly it matched my dress. Upon learning that the handbag and all its contents were made from paper, she asked me to teach the fold to her, the embassy staff, and their guests.

Five years later, when I was in England attending a function at Oxford University, I noticed that the woman sitting next to me was holding a handbag exactly like the one I had designed. I asked her where she had learned how to make it, and she replied, "From my friend, who was at the British Embassy in Rome when some American lady gave an impromptu paper folding lesson"!

Origami crown folded from recycled magazine pages. Created by Laura Kruskal.

In 1996, during my husband's and my third visit to India, we stayed in Pondicherry, where my husband was teaching a three-week course for graduate and postdoctoral students sponsored by The International Centre for Pure and Applied Mathematics. During our visit I created a paper fold that gives me enormous pleasure: a memorial wreath for my husband's mother, Lillian Kruskal Oppenheimer. Grandma Lillian was the founder and director of the Origami Center of America, from which emerged the Friends of the Origami Center of America, now known as Origami USA. She was an amazing woman who did more than anyone to further and encourage interest in origami worldwide. Grandma Lillian died in 1992, three months before her ninety-fourth birthday. Each year on the anniversary of her death I create a new wreath in her memory. It was in Pondicherry, India, that the idea for my 1996 wreath was born.

Each morning I awoke at 5:30 to the chanting of the Surya Pranam Mantra, a prayer that greets the arrival of the sun, followed by the raga Bhairavi, music of the dawn. One morning, unable to sleep, I went out to my balcony to watch the sun bursting through the clouds over the Bay of Bengal. The grandeur of this splendid sight inspired me to create Grandma Lillian's memorial wreath, which I call *Sunburst Over the Bay of Bengal.* The wreath is about two feet in diameter and consists of thirty-eight modules made from $8\frac{1}{2}$-by-11-inch paper.

In an instructional video made during my visit to show my methods and to aid teachers of blind, deaf, and mentally retarded students, some students are filmed working together to interlock the wreath's modules. A model the size of *Sunburst Over the Bay of Bengal* is perfect for a class project, because the modules are easy to make and large enough to handle and assemble without difficulty. I believe that this model is particularly appropriate for Grandma Lillian, who wanted as many people as possible to be involved in origami.

Origami is a lively and enjoyable activity. My intent when teaching origami is to capture the interest of my students and to demonstrate that origami can be beneficial and fun for all. I pride myself on successfully teaching origami to all kinds of people: able-bodied, hearing and visually impaired, infirm, emotionally disturbed, mentally deficient, or physically handicapped. My fervent hope is that origami will become increasingly available for people with special needs.

ACTIVITY

16

Free Folding More Polyhedra from Star-Building Units

If you have completed all of the activities in this book that use star-building units, then you have created models using 3, 6, 9, 12, and 30 units. Did you notice that for each of the models you used a number of units that is a multiple of 3? Because you always use 3 units to create a pyramid, the total number of units needed for any star-building-unit polyhedron will always be a multiple of 3. You can create models using 15, 18, and 21 units. You can even build a model using more than 30 units! And sometimes you can build more than one model using the same number of units. With 12 units you can build a stellated octahedron, and you can also build a model showing three colliding cubes. In this activity you will create your own special models using any number of star-building units, as long as you choose a multiple of 3. When you are constructing your model, be sure to think about color symmetries.

MATERIALS NEEDED FOR EACH STUDENT OR GROUP OF STUDENTS

Squares of origami or patty paper in a variety of colors
Origami journal

GROUPING

Work with a partner or a group. You can each make your own polyhedra, or you can make them with a partner or the group.

EXPLORING YOUR MODEL

Record your answers to the exploring questions in your origami journal. Share your results with others in your group and with the rest of your classmates.

1. Name your polyhedra. Keep a list of polyhedra you make and how many units they use.
2. Are any of your polyhedra symmetrical? Describe the symmetry.
3. Are any of your polyhedra regular solids or related to regular solids? Explain.
4. Categorize your polyhedra in other ways. You can use lists, Venn diagrams, graphs, or other methods of categorization.

Three views of a model made with 12 star-building units

Two views of a model made with 18 star-building units

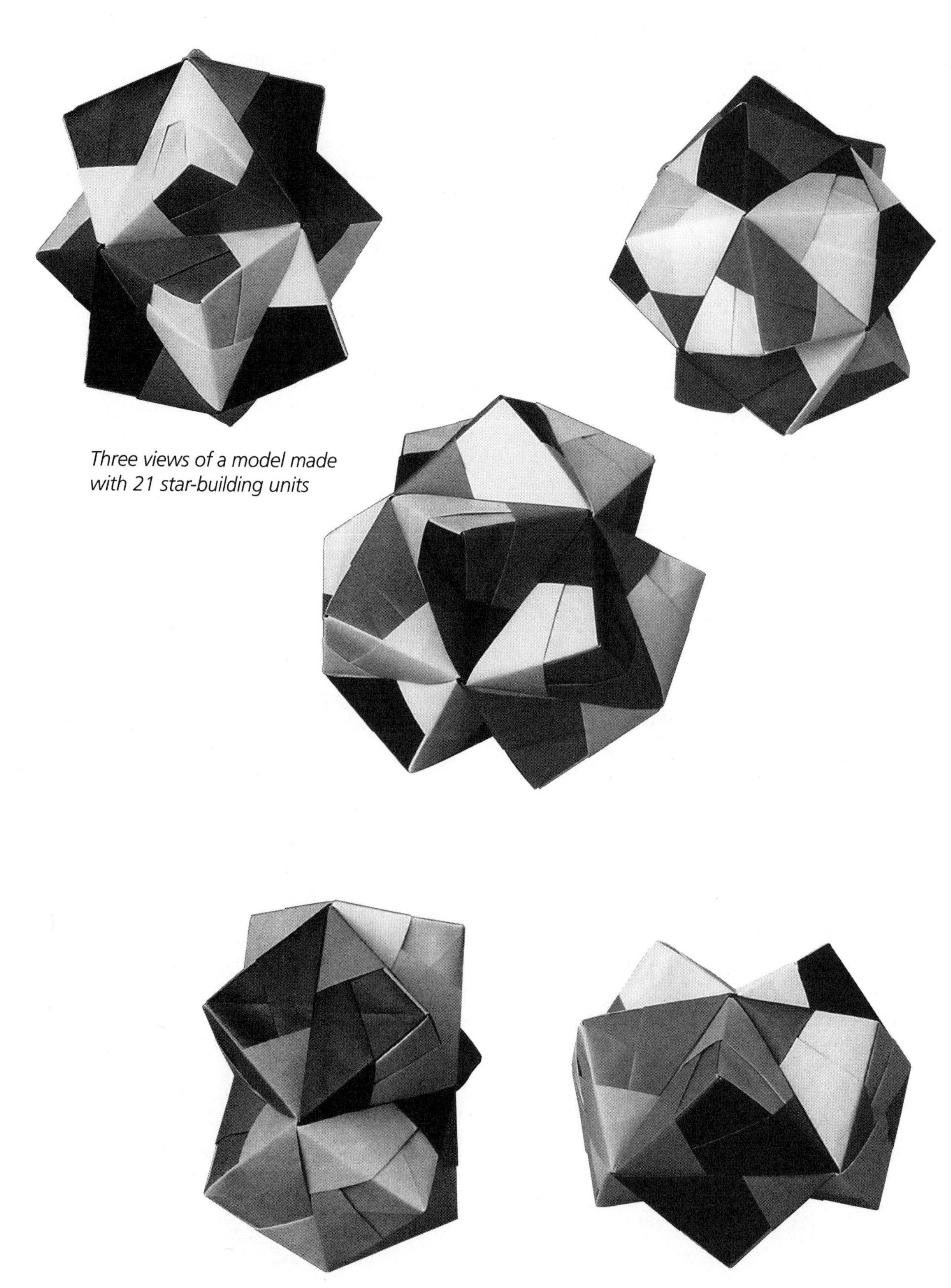

Three views of a model made with 21 star-building units

Two views of a model made with 15 star-building units

ACTIVITY 16

Free Folding More Polyhedra from Star-Building Units

OBJECTIVES

To freely explore and create polyhedra from multiple star-building units
To categorize resulting polyhedra in different ways

MATERIALS NEEDED FOR EACH STUDENT OR GROUP OF STUDENTS

Squares of origami or patty paper in a variety of colors
Student worksheet pages—one copy for each group of students
Origami journal

TEACHER MATERIALS

None

TIME

One or two class periods

GROUPING

Students should work with partners or in groups. They can make individual or group polyhedra.

GENERAL INSTRUCTIONS

Encourage students to be creative and explore possibilities in this activity. You may want to assign this activity as a project that students can work on at home. Be sure to have students share their results. Perhaps you could create a star-building unit model display in your classroom or elsewhere in the school.

ANSWERS

Exploring Your Model

1. Keep a class record of polyhedra the class creates and the number of units required for each. Some students may be very ambitious and go way beyond thirty units. Assemblies of multiples of three units produce interesting polyhedra.
2. Answers will vary. Many of the results will be "relatives" of the stellated octahedron and the colliding cubes.
3. Only the stellated octahedron (eight units) and the stellated icosahedron (thirty units) are related to regular polyhedra. Other models are usually compounds of cubes and pyramids.
4. Polyhedra can be sorted and categorized in at least three ways: whether they are convex or nonconvex, whether or not they are related to a regular solid, and whether or not they have reflection, rotational, or reflection and rotational symmetry.

Glossary

ntiprism A polyhedron that is made up of wo congruent, parallel faces (*bases*) that are onnected by triangular faces (*lateral faces*). An ntiprism is semiregular if all the faces are regular olygons.

pex (of a triangle) The vertex of a triangle that s opposite the side chosen as the base.

Archimedean semiregular solid One of hirteen semiregular solids believed to have been lescribed by Archimedes. These solids include he truncations of the regular solids, the uboctahedron and its relatives, and the cosidodecahedron and its relatives. These hirteen solids, along with the prisms and ntiprisms, make up the set of semiregular solids.

rea of a cube $A = s^2$, where s is the length of a side.

rea of a triangle $A = \frac{1}{2} bh$, where b is the ength of the base and h is the height.

Aristotle A Greek philosopher, educator, and scientist who lived from 384 B.C. to 322 B.C.

bisect To divide into two equal parts.

central angle (of a polygon) An angle whose vertex is the center of the polygon and whose sides extend from the center of the polygon to consecutive vertices.

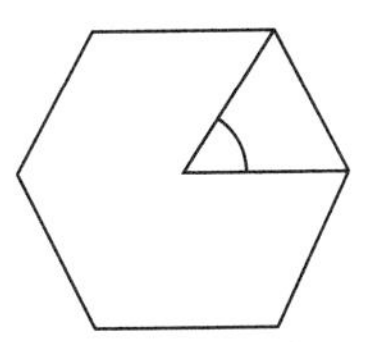

concurrent Lines, segments, or rays that lie in the same plane and intersect at a single point.

congruent angles Angles having the same measure.

congruent polygons Polygons having the same size and shape.

convex polygon A polygon for which any diagonal lies on or inside the polygon.

convex polyhedron (*pl.* polyhedra) A polyhedron for which any line segment connecting two points on the surface of the polyhedron lies on or inside the polyhedron.

cross-section polygon The polygon that results when a plane intersects a polyhedron.

deltahedron (*pl.* deltahedra) A convex polyhedron whose faces are congruent equilateral triangles. There are eight deltahedra.

diagonal A line segment connecting nonconsecutive vertices of a polygon or polyhedron.

diametrical Opposite.

dihedral angle An angle formed by two faces of a polyhedron that meet at an edge. The dihedral angle is the angle inside the polyhedron.

dodecahedron A polygon with twelve faces.

dual edges of interpenetrating polyhedra Edges of polyhedra that are perpendicular bisectors of each other.

dual polyhedra A pair of polyhedra that have the same number of edges and for which the number of vertices on each polyhedron is equal to the number of faces on the other polyhedron. Example: The cube and the octahedron are duals.

dypyramid A polyhedron that consists of two pyramids connected at their bases. Example: The octahedron is a dypyramid.

edge See **polygon** and **polyhedron.**

equilateral triangle A triangle with three sides of equal length.

Euler's formula $F + V = E + 2$, where F is the number of faces of the polyhedron, V is the number of vertices, and E is the number of edges. Euler's formula is true for all polyhedra except those with holes, tunnels, and ring-shaped faces, and for multiple structures (polyhedra inside polyhedra).

exterior angle (of a polygon) The angle formed when one side of the polygon is extended.

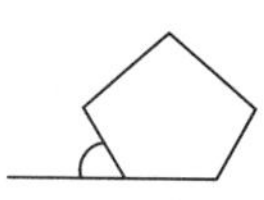

face See **polyhedron.**

Fibonacci series The series 1, 1, 2, 3, 5, 8, 13, 21, 34, . . . Each number in the series is the sum of the two preceding numbers.

Golden Ratio or **Golden Proportion** The ratio created when a line segment is divided into two lengths such that the ratio of the longer length to the shorter length is equal to the ratio of the total length to the longer length. This ratio is approximately equal to 1.618: $\frac{b}{a} = \frac{a+b}{b}$, where a is the shorter length and b is the longer length.

The ratio of the length of the short side to the length of the long side of a **Golden Rectangle** is the Golden Ratio.

Golden Rectangle A beautifully proportioned rectangle in which the ratio of the length of the long side to the length of the short side is the **Golden Ratio** (approximately 1.618).

hexagon A six-sided polygon.

hexahedron (*pl.* hexahedra) A six-sided polyhedron. A cube is a regular hexahedron.

hypotenuse The side of a right triangle that is opposite the right angle. The other sides are called *legs.*

icosahedron A twenty-sided polyhedron.

inscribed polyhedron A polyhedron that is inside another solid, with each vertex of the polyhedron touching the surface of the solid.

interior angle of a polygon The angle formed by two consecutive sides of a polygon.

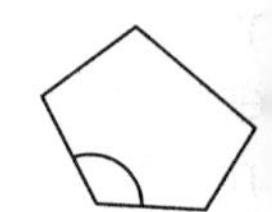

Johannes Kepler A German astronomer and mathematician who lived from 1571 to 1630.

net The result of unfolding and opening a polyhedron so that all faces lie in one plane.

nolid A polyhedron having no volume (as opposed to a solid), made of intersecting congruent polygons. Example: The skeleton of a octahedron is a nolid made of three intersecting squares.

nonconvex polygon A polygon for which at least one diagonal lies outside the polygon.

nonconvex polyhedron A polyhedron for which at least one line segment connecting two points on the surface of the polyhedron lies outside the polyhedron.

octahedron An eight-sided polyhedron.

parallel lines Two or more lines that lie in the same plane and do not intersect.

parallel polygons Polygons that lie in differen planes that do not intersect.

parallelogram A quadrilateral whose opposite sides are parallel.

pentagon A five-sided polygon.

perpendicular bisector of a line segment A line that divides the line segment into two equal parts and is also perpendicular to the line segment.

perpendicular lines Two lines that intersect to form a right angle.

perpendicular planes Two planes that intersect to form a right angle.

Plato A Greek philosopher and educator who lived from approximately 427 B.C. to 347 B.C.

Platonic solids The five regular solids, studied by Plato and the members of his academy.

point symmetry 180° rotational symmetry.

olygon A closed two-dimensional geometric igure formed by connecting line segments ndpoint to endpoint with each segment ntersecting exactly two others. Each line egment is called an *edge* of the polygon. Each ndpoint where the edges meet is called a *vertex* f the polygon.

olyhedral angle An angle formed by three r more intersecting faces of a polyhedron.

olyhedron A solid whose faces are enclosed y polygons. The *faces* are the flat polygonal urfaces. An *edge* is a segment where two faces f the polyhedron intersect. A *vertex* is a point of ntersection of three or more edges.

rism A polyhedron that is made up of two ongruent, parallel faces (*bases*) that are onnected by faces that are parallelograms *lateral faces*). A prism is semiregular if all the aces are regular polygons.

yramid A polyhedron that is made up of one ase that is a polygon and other faces (*lateral aces*) that are triangles. The triangles are formed y segments (*lateral edges*) that connect the ertices of the base to a point (*vertex*) not on the ase.

ythagorean theorem The theorem that tates that for any right triangle with leg lengths a and b and hypotenuse c, $a^2 + b^2 = c^2$.

uasiregular solid A solid whose surface is omposed of two types of polygons, each of which is completely surrounded by the other type f polygon.

rectangle A parallelogram with equal angles.

reflection symmetry Symmetry in which a figure can be reflected about some line of reflection in such a way that the resulting image coincides with the original figure. Also called *line symmetry* or *mirror symmetry.* The line of reflection is called the *axis of symmetry* or the *line of symmetry.* For a polyhedron, there may be a *plane of symmetry.*

regular polygon A polygon that has equal sides and equal angles.

regular solid A polyhedron whose faces are congruent regular polygons and whose vertices all look the same (the same number of faces and edges meet at each vertex in the same way). Regular solids are also called *Platonic solids* and *regular polyhedra.*

rhombus A parallelogram with equal sides.

right isosceles triangle A triangle with a 90° angle and at least two equal sides.

right scalene triangle A triangle with a 90° angle and with three sides of different lengths.

rotational symmetry Symmetry in which a figure can be rotated $n°$ about a point ($n < 360°$) such that the resulting image coincides with the original figure. The figure is said to have a rotational symmetry of $n°$. For polyhedra, the point becomes an *axis of rotational symmetry.* If y is the number of times the polygon (or polyhedron) is rotated $n°$ in order to return to its original position, the polygon (or polyhedron) is said to have *y-fold rotational symmetry*.

semiregular solid A polyhedron whose faces consist of more than one regular polygon and whose vertices all look the same. There are an infinite number of prisms and antiprisms in this set of polyhedra and thirteen Archimedean solids.

similar polygons Polygons having the same shape.

skew lines Nonintersecting, nonparallel lines.

snub As in snub-nosed. *Snub* is *simus* in Latin, which means *squashed.* The semiregular solids include the snub cuboctahedron and the snub icosidodecahedron.

solid A body that occupies three-dimensional space and is bounded by a closed surface. This surface can be curved and/or planar. Examples: The sphere, cylinder, and cube are solids.

space filling Polyhedra that fill up space without gaps, either alone or in combination with other polyhedra.

square A parallelogram with equal sides and equal angles.

stellated Star-shaped. *Stellation* of a polyhedron means that all the faces of the polyhedron are extended until they intersect.

tessellation An arrangement of closed shapes that completely covers the plane without overlapping and without leaving gaps. A *pure* or *regular tessellation* involves a polygon that tessellates by itself. A *semiregular tessellation* is one in which the same combination of regular polygons meet in the same order at each vertex.

tetrahedron A four-sided polyhedron.

trapezoid A quadrilateral with exactly one pair of parallel sides called its *bases*.

truncated polyhedron A polyhedron on which the corners have been sliced off to form new polygonal faces. Five of the semiregular solids are truncated versions of the regular solid In these cases, the corners have been sliced off form new regular polygonal faces.

valency The number of edges meeting at a vertex of a polyhedron.

vertex See **polygon** and **polyhedron.**

volume of a cube $V = s^3$, where s is the leng of a side.

volume of a pyramid $V = \frac{1}{3} \times$ area of base $\times$ height.

Bibliography and Resources

Note: Origami paper is available in packs of one hundred single-color sheets through Kakubundo, 100 North Beretania Street, Honolulu, Hawaii 96817, or in packs of four hundred assorted-color sheets through Key Curriculum Press (1-800-995-MATH). Tomoko Fusè's books are also available through many domestic distributors, including Key Curriculum Press.

Japanese books on origami may be purchased from Origami USA, 15 West 77th Street, New York, NY 10024-5192, or Kinokuniya Books, 10 West 49th Street, Rockefeller Plaza, New York, NY 10020.

Ansill, Jay. *Lifestyle Origami.* New York: Running Heads, 1992.

Includes modules that form deltahedra, pentagonal dodecahedron, and star system polyhedra.

Fetter, Ann E., Cynthia Schmalzried, Nancy Eckert, Doris Schattschneider, and Eugene Klotz. *The Platonic Solids Activity Book* (Visual Geometry Project). Berkeley, CA: Key Curriculum Press, 1991.

A well-organized book containing stand-alone lessons and independent student projects that enable students to explore interesting aspects of the Platonic solids while building three-dimensional models. Prepackaged model-building kits are also available.

Fusè, Tomoko. *Joyful Origami Boxes.* New York: Japan Publications, 1997.

This book teaches you to create beautiful boxes with lids in a variety of polygonal shapes. This book even includes a hepatong box as well as instructions for folding nested boxes.

———. *Origami Boxes.* New York: Japan Publications, 1989.

In this book you will learn how to create beautiful boxes with lids in a variety of polygonal shapes—triangular, square, pentagonal, hexagonal, and octagonal.

———. *Quick and Easy Origami Boxes.* New York: Japan Publications, 1994.

A kit consisting of a sixty-page full-color book with instructions for twenty-two box designs and three packs of origami paper.

———. *Unit Origami.* New York: Japan Publications, 1990.

This is a comprehensive book that is completely devoted to unit origami and shows how to fold and assemble many fascinating polyhedra. It is a perfect complement to *Unfolding Mathematics with Unit Origami.* In fact, students are referred to figures in Fusè's book for some of the lesson extensions.

Hilton, Peter, and Jean Pedersen. *Build Your Own Polyhedra.* Menlo Park, CA: Addison-Wesley, 1988.

Although not really unit origami, the beautiful polyhedra in this book are created using repeated folded *strips.* The text is a valuable resource for polyhedra in general.

Holden, Alan. *Shapes, Space, and Symmetry.* New York: Dover Publications, 1971.

An invaluable source on truncation, stellation, and dual polyhedra. Holden's explanations are particularly clear, and the two-toned models in the photographs are very educational.

Kasahara, Kunihiko. *Origami Omnibus.* New York: Japan Publications, 1988.

The ultimate origami book, this is a huge work with folds for everything from animals to masks to polyhedra. One long chapter is devoted to constructing polyhedra using unit origami. Another chapter, called "Origami to Make You Think," discusses the mathematics of many folds. Mathematical relationships—area, ratio, square root—are highlighted throughout. Some of the extensions in *Unfolding Mathematics with Unit Origami* involve figures from this book.

Kasahara, Kunihiko, and Toshie Takahama. *Origami for the Connoisseur.* New York: Japan Publications, 1987.

The cover states, "not for novices, this is a devotee's origami book, full of challenging forays into the realm of paper folding." One of the few books in English that not only covers complex unit origami and the Kawasaki "iso-area" folding method but also includes Haga single-sheet polyhedra and a host of difficult models of origami animals.

Kenneway, Eric. *Complete Origami.* New York: Japan Publications, 1987.

A compendium of origami from A to Z. There's a neat unit for the pentagonal dodecahedron under "M" for "modular origami." This book even contains directions for folding both flat and ring flexagons.

Neale, Robert, and Thomas Hull. *Origami, Plain and Simple.* New York: St. Martin's Press, 1994.

Many consider this the ideal first book in origami. The book begins with folding basics and progresses gradually to more difficult folds. More than 400 line illustrations are featured, and 30 engaging projects include "talking" birds, fish, and an entire chess set.

Origami USA. *Annual Collection.* New York: Origami USA. Various years.

The *Annual Collection* is published each year for the Origami USA convention held in New York City. The $8\frac{1}{2}$ in. x 11 in. volume usually contains 300+ pages of model diagrams. There are usually a number of geometric models included. For information on collections still available and membership, write Origami USA, 15 West 77th St., New York, NY 10024-5192.

Pearce, Peter, and Susan Pearce. *Polyhedra Primer.* Palo Alto, CA: Dale Seymour Publications, 1978.

This resource is a must. A very accessible resource of definitions and basic information about polyhedra, with simple, clear drawings. The drawings of multiple polyhedra filling space are particularly interesting.

edersen, Jean. "Some Isonemal Fabrics on olyhedral Surfaces," in *The Geometric Vein: The oxeter Festschrift,* edited by Chandler Davis et l., pp. 99–122. New York: Springer-Verlag, 981.

The color plates in this article depict incredible models made by weaving unit strips into polyhedra. The stellation of the icosahedron and of the hexecontrahedron can be modeled with the basic Japanese parallelogram unit.

ugh, Anthony. *Polyhedra: A Visual Approach.* erkeley: University of California Press, 1976.

This comprehensive book offers both basic explanations and in-depth discussions of topics related to polyhedra. Pugh addresses many subtle and fascinating geometric topics that other texts barely mention.

ow, T. Sundara. *Geometric Exercises in Paper olding.* New York: Dover Publications, 1966.

A well-known collection of exercises for folding polygons using one square sheet of paper. Row's explanations are very mathematical.

erra, Michael. *Discovering Geometry: An nductive Approach,* 2nd edition. Berkeley, CA: ey Curriculum Press, 1998.

Serra allows students to truly discover geometry in his lively, complete textbook. The chapters "Line and Angle Properties," "Triangle Properties," "Polygon Properties," "Transformations and Tessellations," "Area," and "Volume" are relevant to the topics in *Unfolding Mathematics with Unit Origami.*

Seymour, Dale, and Jill Britton. *Introduction to Tessellations.* Palo Alto, CA: Dale Seymour Publications, 1989.

An introductory book about tessellations that gives well-organized, detailed explanations of pure and semiregular tessellations; symmetry and transformations, and techniques for generating intricate tessellations. A book of blackline masters is also available.

Stonerod, David. *Puzzles in Space.* Hayward, CA: Activity Resources, 1983.

This charming book offers a series of interesting puzzles related to polyhedra. Blackline masters of nets are included.

Wenninger, Magnus J. *Polyhedron Models for the Classroom.* Reston, VA: National Council of Teachers of Mathematics, 1975.

A very thorough presentation of nets for constructing polyhedra, from the regular solids on up to very complex stellations of polyhedra.

Yamaguchi, Makoto. *Kusudama: Ball Origami.* New York: Japan Publications, 1991.

Kusudama translated from the Japanese means "decorative ball," but in reality it is a decorative polyhedron. This English translation of Yamaguchi's work explains the construction of these polyhedra via unit origami methods.